PERSONAL COMPUTERS
IN CHEMISTRY

Edited by
PETER LYKOS
Illinois Institute of Technology

PERSONAL COMPUTERS
IN CHEMISTRY

A Wiley-Interscience Publication

JOHN WILEY & SONS, New York • Chichester • Brisbane • Toronto

Library of Congress Cataloging in Publication Data:
Main entry under title:

Personal computers in chemistry.

"A Wiley Interscience publication."
"Papers presented at the August 1979 national
American Chemical Society meeting held in Washing-
ton, D.C."
    Includes index.
    1. Chemistry--Data processing--Congresses.
2. Microcomputers--Congresses. I. Lykos,
Peter. II. American Chemical Society.
QD39.3.E46P47   542'.8   80-25445
ISBN 0-471-08508-1

Printed in the United States of America

10 9 8 7 6 5 4 3 2

PREFACE

The cost of personal computers is now sufficiently low that anyone can afford to purchase one. Furthermore, their level of sophistication in terms of human-engineered operating systems, software utilities, interfaces, and industrywide standardization is such that responsible chemists can now incorporate these powerful and flexible information processors into their professional activity, whether that be management, teaching, writing, or research.

A carefully selected subset of the papers presented at the August 1979 National American Chemical Society Meeting held in Washington, D.C., has been edited and indexed for this volume. The papers represent a snapshot in time of the range and depth of use and application of microprocessor-based computer systems designed to be affordable by individual users. Indeed the theme of the symposium was "Chemists and Computers: Now One on One."

These 20 papers were selected out of the more than 40 presented so that, within the constraints of space, a balanced and representative range of aspects of personal computers in chemistry would be available to the reader. The papers can be grouped by emphasis as follows:

1. The first group of five papers, from Koontz and Benezra to Gray and Workman, represents the most immediate and direct use of microprocessor-based systems. These papers deal with the use of a single small computer to collect data from and/or to control a single chemical instrument. Two of the applications are from industrial research laboratories, three from universities. They touch on the important considerations of cost,

reliability, and maintenance. In addition the important role of graphics display, even in these relatively straightforward applications, is highlighted. Also a wide range of data rates is accommodated.

2. The next three papers, from Beavers et al. to Fischer et al. go one step farther to deal with supporting several instruments, via a "star" network, with a central processor. The star networks range from using a central processor only to using microprocessors, one at each experiment, as an "intelligent" interface to a conventional time-sharing system. The additional complication of contention for limited resources is addressed here, as well as the positive aspects of having the additional flexibility inherent in a more complex system.

3. The three papers of which Kahn is author or coauthor were included as a set to give a case history of bringing a minicomputer into a laboratory complex, with a variety of instruments to be served, and thus to illustrate the rapid assimilation and increase in information processing sophistication from practically zero (in real time computing) to a comfortable and accepting environment, with very limited hardware and human resources, all over a short time span. Here the minicomputer was introduced first and the appetite for and ability to use microprocessors came later.

4. Pollnow's paper is a harbinger of things to come in terms of a stand-alone device of modest cost, high resolution graphics display, and a comprehensive set of "canned" programs that, as a system, enables the chemist to use, easily and conveniently, a variety of numerical mathematical tools. Again the importance of a well engineered human/machine interface is being illustrated (ergonomically designed terminals constitute a facet of that thrust).

5. "Talking microcomputers" is the focus of the next paper by Morrison et al. Both speech recognition for input and speech generation for output are capabilities now available in the personal computer cost and ease of use range. This paper describes a system designed to aid blind students in chemistry laboratories, but the range of application for sighted persons is broad and growing.

6. The personal computer in chemical education represents a rapidly growing application. Indeed the American Chemical Society's Committee on Professional Training is conducting a comprehensive survey of the 550 chemistry departments that can award ACS-certified bachelor degrees in order to determine how to revise the certification criteria to take that into account. Only three papers were selected to represent this area: Smith's paper on the adaptation of the PLATO computer-assisted-instruction system to a stand-alone device using a microprocessor; Brubaker's paper, which distills a considerable amount of experience using the more popular personal computers to support a large number of short programs somewhere between tutorial drill and practice and a full blown (entire course) based on CAI; and

Burden's paper, which is concerned with teaching computer/instrument interfacing in a liberal arts college using a team design approach.

7. The paper by Macero et al. is included because it gives an excellent survey of the range of personal computer processors, peripherals, operating systems, and higher level languages available. A useful supplement to this paper are the three monthly publications *Creative Computing, Byte Magazine,* and *Computer Design,* where the interested chemist using microprocessor-based systems of the personal computer genre can keep abreast of new developments and can share experiences with other users.

8. The last three papers represent the phenomenon of chemists' having gotten so deeply into computer system design that they may be called computer scientists or engineers as well as chemists. The first, by Schilling, goes into a multidisciplinary approach to the development of an intelligent terminal designed around the needs of students in chemistry, engineering science, and computer science. The second, by Woodward, gives a survey of the important features characteristic of an interactive laboratory data system. That cumulative wisdom comes from an extensive computers-in-the-chemistry-laboratory activity that has evolved over many years at the University of North Carolina at Chapel Hill. The third paper represents a forward-looking attempt to capitalize on the reality of mass-produced very large integrated circuits that are complete and powerful central processing units. It may be the case that chemists will learn how to assemble their own multiprocessor powerful computers, including so-called number crunchers, from easily interfaceable modules. The Neece-Ostlund paper reveals some of the basic considerations involved.

What will be done in the future regarding development and use of microcomputers? It has been said that we tend to overestimate what will happen in one year and to underestimate what will happen in five. This is perhaps closer to the truth in chemistry than in any other discipline.

PETER LYKOS

Chicago, Illinois
August 1980

CONTENTS

PERSONAL COMPUTERS
IN CHEMISTRY

Use of the UNC Microcomputer in an Industrial
Analytical Environment:
HPLC Applications

Application of microcomputer technology has resulted in a versatile, high-performance data acquisition and control system for high-performance liquid chromatography (HPLC). The HPLC system was designed around commercially available hardware components and consists of: two high-pressure pumps with voltage programmable flow rates[1], a digitally controlled and sensed microprocessor-based autosampler[2] and a variable wavelength UV detector[3]. In this configuration, the chromatograph is capable of stand-alone as well as computer-controlled operation.

An analysis of the HPLC experiment reveals five distinct phases:

1. Experiment configuration
2. Data acquisition
3. Computation of retention times, peak height, and peak areas
4. Data presentation
5. Archival storage and retrieval

The experiment configuration phase consists of establishing parameters which define the data acquisition per se. This is an interactive period on a human time scale requiring little I/O or computation. The data acquisition is characterized by computer control of system functions and events, as well as digitization of low level (< 1 volt) signals over a wide dynamic range for long periods of time when signal reference (base line) values are subject to drift and changes in solvent composition. Determination of retention times and peak areas is computation, rather than I/O, intensive, but usually brief when compared to the data acquisition phase. Data presentation is an interactive period of graphical display utilizing moderate levels of calculation (such as integration or autoscaling) to ultimately produce a time domain chromatogram for interpretation. Storage and retrieval of raw data and peak parameters for future reference or reprocessing requires access to a mass storage peripheral, but little or no computational power. From this analysis, it was determined that a properly implemented microprocessor should be sufficient for all phases of the HPLC experiment.

The heart of the chromatographic system is a microcomputer designed and optimized for laboratory data acquisition and instrument control. This implementation of the Intel 8080 processor was originally done at the University of North Carolina at Chapel Hill by Woodward, Reilley, et al.[4] and is commercially available[5] although the particular unit used in this work was constructed in-house.

Choice of the UNC microcomputer system over other mini- and microcomputers was based on several criteria:

1. Direct Memory Access

    The UNC implementation provides easy to use DMA hardware in support of mass peripherals, data acquisition and bootstrap operations at rates up to 667 KHz.

2. Dynamic Random Access Memory

    Dynamic RAM is low in power consumption (proportional to access rate rather than quantity as with static components). Fewer packages are required due to higher bit densities for dynamic components. Dynamic RAM is typically less expensive on a per bit basis than static equivalents. Refresh circuitry is provided on the CPU board.

3. Synchronous System Bus

    All bus activity is synchronous (including memory refresh), which allows utilization of software timing loops where accuracy is desired.

4. Video Display

    8K bytes of memory are utilized to form a 256 x 256 bit-mapped graphics display on a consumer grade television or standard video monitor.

5.   High Performance Peripherals

    a.   16-Bit Integrating ADC and 20 Hz Clock

    b.   Audio Cassette Interface

    c.   Quad 12-Bit DAC

    d.   Digital I/O Interface

6.   Low Cost

    Total microcomputer system component cost for in-house construction was under $2,000.

Figure 1 illustrates the data system interface to the chromatograph. Currently, only one analog signal from the variable wavelength detector is available, however, channels for electrochemical and RI detectors are included. Analog control of the pumps requires two DAC channels configured for 0 to 5 volt ranges which correspond to flow rates from 0 to 10 ml/min. Digital control and sensing of the autoinjector is via TTL compatible I/O ports on the digital interface. Figure 2 shows the microcomputer configuration.

The interactive software is written in BASIC and supports all hardware functions including: pump control; sample injection; data acquisition; system timing; video and hardcopy graphics; hardcopy printout; and named program and data files on cassette tape. Table 1 illustrates the use of the "DAC" statement to control the pumps. A parenthetically enclosed argument following the statement indicates which channel is to receive the analog output. The value assigned to the specified channel is in the range of 0.0 to 1.0 where 0.0 corresponds to the lower DAC bound and 1.0 corresponds to the upper DAC bound.

Table 1
Pump Control from BASIC

```
10 DAC(2) = .1; REM SET CHANNEL 2 PUMP TO 1 ML/MIN
20 DAC(3) = B + M * LOG (X*Y/Z); REM SET CHANNEL 3 PUMP
```

Table 2 shows a typical data logging routine utilizing the "IADC" function to access the 16-bit integrating ADC. A parenthetically enclosed argument following the function indicates which channel is to be converted. Following the channel specification is an optional variable which is incremented by the

Figure 1: Data System Interface

Figure 2: Microcomputer Configuration

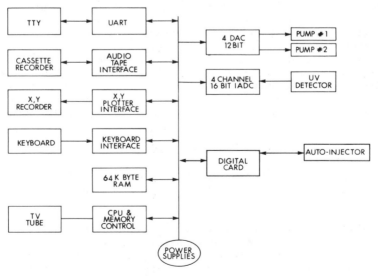

4

time in seconds since the ADC was last read. The particular example shown averages N conversions for D seconds and displays the mean on the video display in real time. Control of the autoinjector is via the ubiquitous "PEEK" and "POKE" functions.

Table 2
Data Acquisition in BASIC using the Integrating ADC

```
10    D = 0.5; T = 0
20    X = IADC(1)S; S = 0
30    X = 0; N = 0
40    X = X + IADC(1)S
50    N = N + 1
60    IF S < T THEN 40
70    PRINT X/N
80    T = T + D
90    GOTO 30
```

Table 3 outlines the use of the video and hardcopy graphics commands. Three commands control vector and character generation on the video display while a single command provides these same functions for an XY plotter driven from two DAC channels with pen control implemented.

In BASIC, the video screen is organized as a 256 x 256 dot matrix. Each pixel is specified by a pair of screen coordinants with (0.0, 0.0) being the lower leftmost element and (1.0, 1.0) being the upper rightmost element.

The "PNT" statement generates a single point in background complement at the parenthetically delimited screen coordinants. The "VCT" statement is similar to the "PNT" statement except that instead of generating a point, the best straight line is interpolated from the last specified screen coordinants to the new coordinants specified in the "VCT" statement. Annotation of video graphic displays is via the "NOTE" statement. The output string generated by the BASIC interpreter is positioned on the video display at the specified coordinants.

The "PLOT" statement accesses an XY plotter interfaced to DAC channels 0 and 1. The pen is raised or lowered as indicated and moved to the specified coordinants. If the pen is down, the best straight line is interpolated to the new coordinants. If an optional output string is specified, it will be printed at the new coordinants as with the video "NOTE" statement.

Table 3
Graphics Generation from BASIC

10	PNT (.5,.5)	generates a single point at the center of the video display.
20	VCT (1,1)	generates a vector from the current location to the upper righthand corner of the video display.
30	NOTE (X,Y) "HELLO"	writes the text string to the video display beginning at the location X,Y
40	PLOT (0,X,Y)	slews the plotter pen to X,Y
50	PLOT (1,0,0)	interpolates a vector from the current location to the lower lefthand corner of the plotter area.

Experimental. Application of HPLC gradient programming to production and quality assurance analyses is limited by the reproducibility of the gradient. Precise gradient generation, until recently, has been at best difficult and, most certainly, expensive. The experimental system for this work generates gradients by means of two independent pumps which are programmed by the microcomputer in lieu of a proportional controller. This open loop approach relies on the accuracy of the control voltages produced by the microcomputer and the stability of the internal pump electronics. High-pressure mixing avoids the necessity of synchronizing solvent proportionation with the fill stroke of the pump and thus allows utilization of less complex and, consequently, less expensive pumps.

In order to determine the controllable range of the pumps, a series of small step gradients were applied to a water solvent system where the "B" solvent was doped with acetone. Figure 3 shows the theoretical and empirical responses for this program. Note that at 6% B (0.09 ml/min) the measured response is within 0.2% (absolute) of theoretical. Below this flow rate the pumps are not analytically controllable. This can be improved by substituting a 1/3 speed motor option in the pump.

Figure 3: Small Step Response

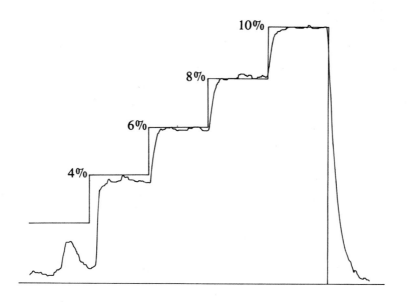

Figure 4: Continuous Gradient Response

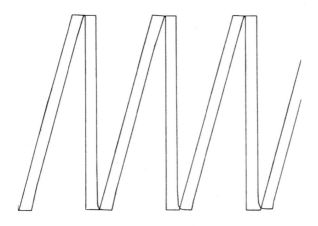

The response of the pumps to a continuous program is illustrated in Figure 4. Analysis of these curves show that the pump output tracks the control voltage so that the total flow rate is constant except when either pump flow is between 0 and 0.1 ml/min.

To demonstrate gradient reproducibility, a mixture of benzene, naphthalene, anthracene and biphenyl was separated using the solvent profile in Figure 5.

Figure 5: Four-Component Separation

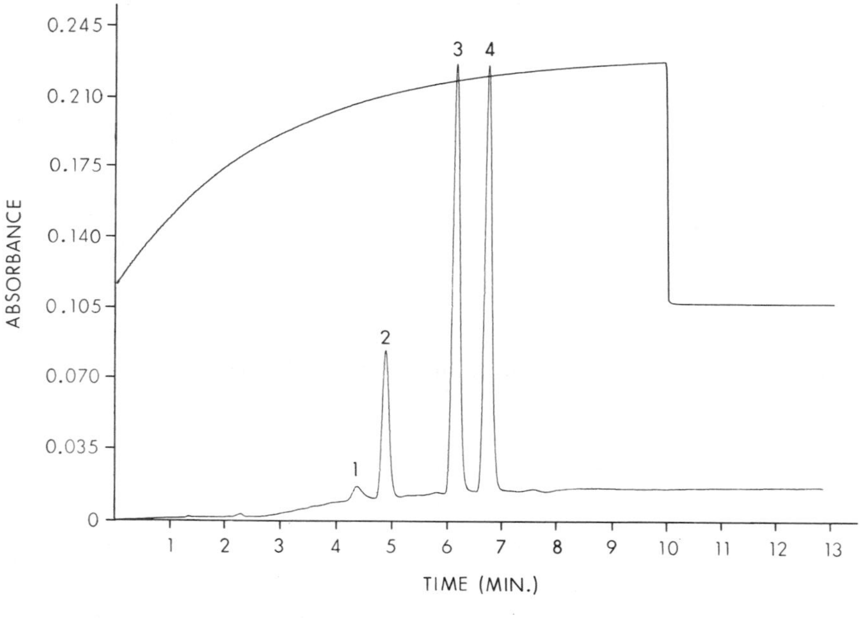

Column: 25cm × 4.2mm Spherisorb ODS-GP
Mobile Phase: Solvent A-H2O
              Solvent B-Methanol
Gradient: 50% to 100% B, Concave
Flow Rate: 1.5ml/min
Detector: UV/254nm
Peak Identity: 1. Benzene
               2. Naphthalene
               3. Biphenyl
               4. Anthracene

The results for 10 separations, given in Table 4, show that retention times are reproducible to better than 1%.

Table 4
Replicate Retention Times for a Four-Component Mixture

Benzene	Naphthalene	Biphenyl	Anthracene
277.5	300.5	376.0	411.0
273.5	300.5	375.5	410.0
273.0	301.0	375.5	410.5
271.0	301.0	376.0	411.0
271.5	301.0	375.5	409.5
271.5	301.5	376.0	410.5
271.5	301.5	376.0	411.0
272.0	303.0	377.0	411.5
271.0	302.0	376.0	410.5
271.0	302.0	376.5	411.5
Avg. 272.4	301.4	376.0	410.7
S.D. ±2.0	±0.77	±0.47	±0.63
%RSD 0.73%	0.26%	0.13%	0.15%

Often it is desirable to perform isocratic separations particularly in the production or quality assurance environment. Isocratic separations require less sophisticated equipment and thereby reduce the cost per analysis. To determine if an isocratic separation is feasible, solvent scouting methods are often employed to optimize the solvent composition for speed and separability. This process has been automated by using an autosampler to replicate sample injections and the microcomputer to generate a series of isocratic solvent compositions. Figure 6 shows the results of such a solvent scouting run. A mixture of aniline, N-methylaniline, and N,N-dimethylaniline was injected repeatedly each time increasing the percentage of water in the methanol solvent. Unattended experiments like these can quickly ascertain the feasibility of isocratic separations and thereby eliminate the need for more costly gradient equipment.

Conclusions. Table 5 shows the component cost of the system. Note that the cost of the microcomputer is only about 10% of the total system. Addition of this much data acquisition and computation capability can nearly double the cost of commercial systems, if it can be done at all.

Figure 6: Solvent Scouting

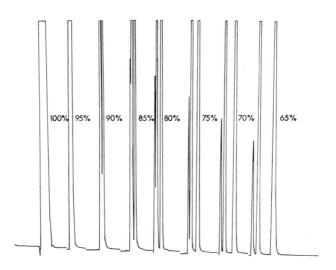

100%	95%	90%	85%	80%	75%	70%	65%

Table 5
Cost of HPLC System

2 pumps (LDC Constametric II)	6,000
Micromeritics Model 725 Autosampler	5,600
LDC Spectromonitor III UV Detector	4,300
Recorder	900
X-Y Plotter (Hewlett-Packard 7010B)	1,500
Microcomputer	2,000
Total	$20,300

Unlike academic environments, hardware cost does not account for the largest portion of expense for an analysis in industry. Certainly, the cost savings in manpower far exceed those for hardware to implement an automated analysis.

The key element in an automated analysis is usually a computer. The microcomputer selected for this work is flexible and easy to use. Modification of system parameters, development of new methods, and system reconfiguration are facilitated by the hardware modularity and stand-alone software for the UNC microcomputer. Since all programming is done in BASIC and no specialized software development system or ROM programmers are required, individual scientists can make software changes without intervention of a software support group. The modular hardware organization provides a convenient means for addition or reconfiguration of equipment and, also, for maintenance. The low cost of microcomputer components allows a full set of boards to be on the shelf without a large capital investment in spare parts. In-house maintenance is now reduced to isolation at the board level so that a quick substitution by non-expert personnel can have the microcomputer back in operation in a matter of minutes. The faulty board can then be repaired by technical staff at their convenience. Such flexibility minimizes the need for and expense of vendor support.

One area where high performance is desirable is mass storage. Even though the UNC audio cassette implementation exceeds other cassette systems in performance, the inherent properties of cassette transports (i.e., tape speed, frequency response, etc.) limit data transfer rates relative to a human time scale. However, this is not a limitation of the microcomputer since other means of mass storage are available: cartridge tape, floppy disk, 8-bit parallel and asynchronous serial compatible devices (e.g., a host computer).

Acknowledgments. The authors gratefully acknowledge Mr. J. O. Hewett for his assistance in construction and installation of the microcomputer and Mr. W. S. Woodward for his continued counsel in matters of microcomputer hardware and software.

Footnotes.

1. LDC Constametric II
2. Micromeritics Model 725 Autoinjector
3. LDC Spectromonitor III Variable Wavelength UV Detector
4. W. S. Woodward and C. N. Reilley, Pure and Applied Chemistry, 50, 785-799 (1978).
5. Digital Specialties, Carrboro, NC 27510

S. C. CRUMPTON, M. S. ADLER,
J. E. RAMIREZ, and R. J. HANRAHAN

CHAPTER 2

Microcomputer-Based Data Acquisition System for Fast Chemical Kinetics

Introduction. The fast kinetic techniques of flash photolysis and pulse radiolysis, developed in the early 1950's and early 1960's respectively, have in common the necessity of collecting a large amount of data in a short period of time.[1] Early work using both techniques emphasized the time regime from 10^{-6} to 10^{-3} seconds, although both techniques subsequently progressed to the nanosecond and picosecond time scales. The sub-nanosecond time scale involves major difficulties, requiring streak-camera or boxcar averaging techniques, and is beyond the scope of this paper.[2]

Early work in flash photolysis and pulse radiolysis, through about 1970, was generally done by photographing an oscilloscope trace, followed by tedious manual digitization of the photographic image, usually on Polaroid film.[1-3] This material, although convenient, is rather expensive. It is not unusual for an active laboratory using the photographic method to consume four to five rolls of film daily, leading to an annual film bill of several thousand dollars.

The incentives of decreasing manual effort and increasing the quality and quantity of data obtained led several laboratories to develop computer-based

methods for the acquisition and numerical analysis of
optical signal vs. time data from flash photolysis and
pulse radiolysis experiments.[4-12] Development of such
equipment was facilitated by the commercial availability of devices called "fast transient digitizers,"
which are capable of acquiring analog signals at rates
as fast as 100 MHz, converting the data to digital
form, storing it in internal fast memory, and retriev-
ing the stored data subsequent to the pulse experiment,
in a form and on a time scale appropriate for further
analysis, using either an oscilloscope, graphic recorder, or data acquisition computer.

   Most of the fast-pulse data acquisition systems
described in the literature were developed at national
laboratories or similar research establishments such
as the Argonne National Laboratory, the Hahn-Meitner
Insitute in Berlin, the Australian Atomic Energy
Research Establishment, etc. Typically, the equipment
used has involved a powerful 16 bit minicomputer such
as the PDP 11, with disks or tape drives, line printers, video display, etc. Total investment in such
equipment could run to $50,000 or more.

   The apparatus which we developed is similar in
concept and function to previously described systems
such as the system originally built at the Mellon
Institute and later duplicated at the Hahn-Meitner
Institute.[5,7] Our facilities differ primarily in that
the equipment is based on "personal computer" type
hardware using the S-100 bus. The total cost was
about $7,000 or roughly 10% of the earlier systems.
There is a tradeoff involved, primarily in computing
time (typically 1 min. per pulse experiment, vs. about
5 sec.) but our unit nearly rivals the more expensive
equipment in versatility and convenience.

   Of the systems described in references 4-12, only
those built by Thornton and Laurence[11] and Barrett et.
al.[12] involve LSI type processors. The former unit is
built around a DEC PDP11/03 which is actually a fairly
powerful minicomputer implemented in LSI format, using
peripherals from the PDP11 family. The latter system
is a minimally configured Altair 880 intended primarily
for pedagogical applications. To our knowledge, a
full-featured fast pulse data acquisition system based
on an 8-bit microprocessor has not previously been
described in the literature.

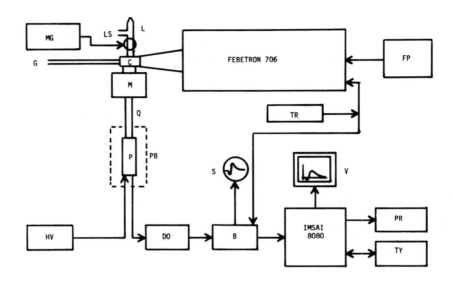

Figure 1 - University of Florida Pulse Radiolysis System

B	- Biomation 610B	MG	- Microwave generator
C	- Reaction chamber	P	- Photomultiplier
DO	- Dual op amp	PB	- Lead cave
FP	- Febetron power supply	PR	- Printer
G	- Sample handling system	Q	- Quartz rod light pipe
HV	- PMT power supply	S	- Oscilloscope
L	- Iodine lamp	TR	- Trigger source
LS	- Lamp flow system	TY	- Teletype
M	- Monochromator	V	- Video

Description of Equipment. Figure 1 shows a block diagram of the pulse radiolysis system and its associated data reduction equipment, located in the Radiation Chemistry Laboratory at the University of Florida. The chief experimental apparatus is a Field Emission Corp (Division of Hewlett Packard) Febetron Model 706 pulsed electron accelerator, capable of delivering a burst of ionizing radiation consisting of 600 keV electrons, with total energy of ca. 12 J, pulse half-width of about 3 ns, and time between pulses of about 1 min. The radiation beam traverses a vacuum-tight aluminum box C, containing an appropriate sample gas to be investigated. Decomposition of the sample subsequent to the pulse is probed using UV-Visible absorption spectrophotometry, involving a microwave glow discharge lamp L (for atomic resonance light), a monochromator M, and a photomultiplier P as shown in the drawing. For the results described below, the monochromator was a 0.25m Jarrel-Ash model 82-415 instrument, and the photomultiplier was an EMI type 9650QB, operated at about 1000 v. Light was conducted from the monochromator to the photomultiplier tube via a bundle of Suprasil quartz rods acting as a light pipe.

Output of the photomultiplier tube was fed via two fast operational amplifiers in series to a Biomation Model 610B transient digitizer. This unit has a memory consisting of 256 six-bit words, and a maximum digitization rate of 10 MHz. The Biomation unit can output its memory contents repetitiously to an oscilloscope, allowing an ordinary oscilloscope such as the Tektronics model 535A which is available in our laboratory to act similar to a storage oscilloscope. The section of the apparatus which is the primary subject of the present paper consists of the portion to the right of the Biomation unit in Figure 1. The computer data reduction system is based on an IMSAI 8080 microcomputer, consisting of control panel, main processor based on the Intel 8080 chip, Processor Technology Co. "3P + S" parallel and serial input-output board, and 48 kbytes of static semiconductor memory. System programs and data can be stored on an 8-inch, hard sectored floppy disk (Innovex Model 410) interfaced by a Peripheral Vision Co. controller board. Good quality visiual display of optical signal vs. time data is provided on a 12-in. video screen using a "Merlin" brand video interface system. An exact copy

of the screen image, which is bit-mapped from computer
memory, can be transferred to an Integral Data Systems
model 225 line printer for permanent record purposes.
The entire data acquisition and reduction sequence is
carried out via operator dialogue entered from a keyboard. A record of these transactions is made on a
Teletype, as indicated in Figure 1.

In order to speed up the numerous calculations
associated with linear least squares treatment of the
data, the system was provided with a NorthStar floating-point hardware board, capable of carrying out
floating point add, subtract, multiply, and divide
operations with times of roughly 10, 20, 50, and 100
micro seconds respectively, on numbers with as many as
14 decimal digits in the mantissa. (For the present
work, the system was operated using 8 decimal digits.)
All items as described were available from commercial
vendors, and were used in our experiment with essentially no modifications. The only custom hardware was
a small interface between the output of the Biomation
and the input to a parallel port on the Processor
Technology 3P + S I/O board. This circuit involves
inversion of the data lines using a 74LS04, handshaking using a 74LS00, and synchronization using a 74LS74
flipflop (Figure 2).

In contrast to the hardware development work,
which involved the direct use of standard, commercially
available components almost exclusively, the software
work involved with the data acquisition project involved a large amount of custom programming, much of it in
assembler language. The software for the Peripheral
Vision disk operating system was used essentially without modification, although it was necessary to reassem-
ble the disk bootstrap program, furnished in the form
of a read-only memory chip, which overlapped the ROM
operating system for the "Merlin" video board. Fortunately, the Peripheral Vision operating system was
thoroughly documented, including source listings.

In developing our system, we made a decision to
avoid undocumented software (that is, software with no
source listings available) as much as possible. This
ruled out the use of most available BASIC interpreters,
since most vendors do not supply source listings.
After examining Lawrence Livermore BASIC and BASIC-E
(public domain, but a compiler rather than an interpreter) and Processor Technology 5-K BASIC, we chose

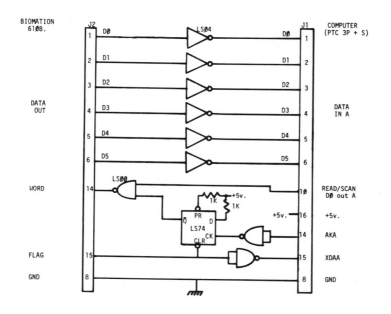

Figure 2.   Interface from Biomation 610B transient digitizer to IMSAI microcomputer using Processor
Technology Co. 3P + S parallel/serial interface board
(shown on right) plus custom circuitry.

the latter primarily because of certain features of
the math package. The math package was subsequently
deleted, however, and replaced with calls to the North
Star floating-point board as described above. It was
necessary to add numerous elaborations to the Processor
Technology 5-K BASIC, including log and exponential
functions, the IN and OUT commands to interact with
the parallel ports, and the PEEK and POKE commands to
examine or change individual memory locations. An
assembly language subroutine CALL command was already
included in the language. The PEEK and POKE commands
furnish both "poke-byte" and "poke-word" versions, the
latter facilitating 16 bit data transfers. Recently,
some primitive string features have been added to the
language, allowing the operator to respond to program-

originated requests for further orders by word-commands rather than numbers.

Further aspects of the software work involved a short machine language routine to transfer data from the Biomation to the IMSAI memory, graphical routines to present visual information on the video screen using features of the "Merlin" system, and further manipulations of graphical data necessary to dump the video image to the Integral Data Systems Printer. The main graphics program is written in PL/M cross-compiler language, and the program to output graphics to the printer is assembler.

The main supervisory program, which allows inter-action with the operator, is written in BASIC. After setting up an experiment and firing the Febetron, a number of commands are available to the operator, as follows:

1) Transfer the data from the Biomation to the computer memory;

2) Display raw data on the video screen;

3) Set left and right cursors to define the region over which data reduction will take place, and set a horizontal cursor to define the long-time base line for immediate curve fitting via taking logarithms, linear least squares reduction, and taking anti-logs to reproduce the experimental curve; (In the most recent version of the software, the "infinite time" baseline is set auto-matically.)

4) Transfer of individual experimental results to a buffer for summing, if desired, in order to improve signal to noise ratio;

5) Display and curve-fitting of data from the buffer;

6) Print-out of the graph of the curve-fitted data on the line printer, using either single-experiment data or the buffer.

Figure 3 is a photograph of the data acquisition com-puter system. Figures 4-7 display typical experimental results and are discussed in the next section.

Typical Results and Discussion. The experiment performed to illustrate operation of the system con-sists of pulse radiolysis of methyl iodide vapor, with pressures in the range of 1-10 torr, using 100-500 torr of Argon buffer gas to aid in absorbing the radiation

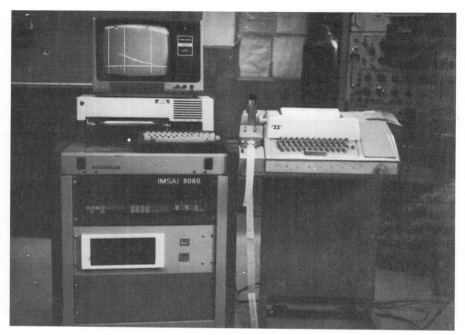

Figure 3.   Photograph of data acquisition system in operation.

beam and transferring energy to the methyl iodide. The formation and decay of excited iodine atom in the $^2P_{1/2}$ state was followed using an atomic iodine resonance lamp, operated at 206 nm. Figure 4 shows raw data from a single pulse experiment, Figure 5 shows the smoothing which results when 3 pulse experiments are performed and the results added in a buffer, and Figure 6 illustrates curve fitting of the buffer shown in Figure 5. Subsequent aspects of the experiment involved determination of the apparent deactivation rate as a function of CH_3I pressure. The apparent rate of deactivation, which was found to be first order in each individual experiment, was interpreted as due to an efficient deactivation process involving methyl iodide superimposed upon a much less efficient deactivation process involving the buffer gas Argon; rate constants of 2.0×10^{-13} and 6.2×10^{-15} cc/molecule second were determined for these two processes, respec-

Curve fit # 2   normal run

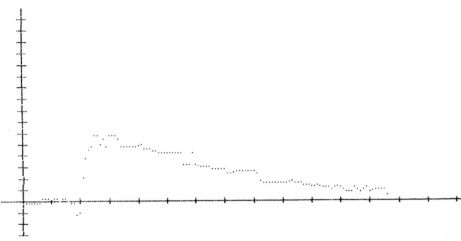

Scale: 4   micro-seconds per X division
        0.0781   volts per Y division (1 volts p-p)

Figure 4 - Graphical display of optical signal versus time for a
single Febetron pulse. Conditions: 3.2 torr methyl iodide,
180 torr Ar.

Curve fit # 4   normal buffer, 3 additions

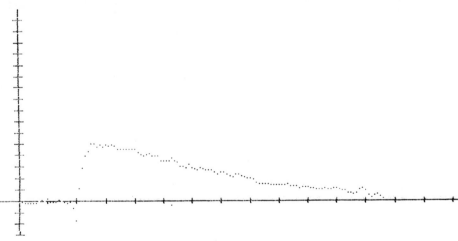

Scale: 4   micro-seconds per X division
        0.0781   volts per Y division (1 volts p-p)

Figure 5 - Graphical display of optical signal versus time for
three Febetron pulses sumed into buffer. Conditions as in Fig. 4.

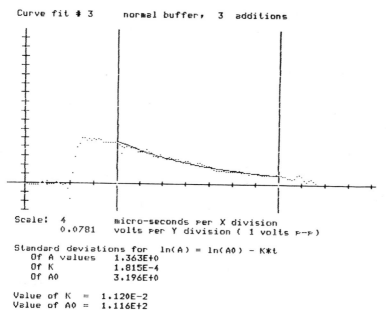

Curve fit # 3   normal buffer, 3 additions

Scale: 4   micro-seconds per X division
        0.0781   volts per Y division (1 volts p-p)

Standard deviations for ln(A) = ln(A0) - K*t
    Of A values   1.363E+0
    Of K   1.815E-4
    Of A0   3.196E+0

Value of K = 1.120E-2
Value of A0 = 1.116E+2

Figure 6 - Curve fit of data from Figure 5.   Rate constant in arbitrary units.

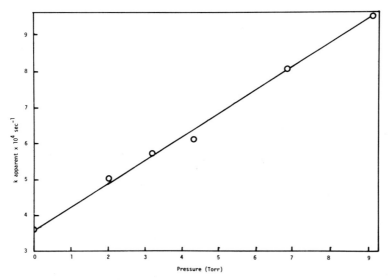

Figure 7 - Pulse Radiolysis of Methyl Iodide With 180 Torr Argon. Deactivation of I· 5 $^2P_{1/2}$ Measured at 206.2 nm.

k_{MeI} = 2.0 x 10^{-13} cc/molec sec ; k_{Ar} = 6.2 x 10^{-15} cc/molec sec

21

tively, based on the slope and intercept of the line
in Figure 7.

The apparatus as described provides for complete
and convenient reduction of data, with first or
second order rate constants (currently in arbitrary
units), error limits, and graphical display of the
results, all produced at the operator's request. It
would require about 0.5 - 1 hour to carry out manual
data measurement on a Polaroid slide, followed by hand
calculation of a best-fit rate constant for each
experiment. The computer performs this work in approximately two minutes depending on the number of points
fitted. Clearly, the availability of the computer
leads to the acquisition of both more data and better
quality data per work day. The system is cost effective even if it is viewed merely as a way to eliminate
the use of instant camera film. The entire investment
in the data acquisition hardware system could be re-
couped in ca. 3 years in savings on film alone.

There are numerous favorable aspects to the use
of "personal computers" for laboratory data acquisition, as seen in the present work. To a considerable
extent, however, the work described here was made
possible by the availability of competent and inexpensive student help, both for building computer kits
and for the necessary software work. Because of the
trend towards decreasing prices in microcomputer hardware, especially in memory chips, it is now possible
to duplicate the hardware of the system using factory
built units, at a price about the same as we paid for
kits. The question of software is a more difficult
one, however, and more significant in the long run.

There are several aspects of our selection of an
S-100 bus computer which deserve special mention.
Early units of this type had some problems with noise
on the bus. This problem has been eliminated in most
of the more recent implementations of the S-100 bus.
Even with an early-generation unit such as our IMSAI
8080, the problem is effectively solved by use of a
plug-in bus-terminator board. Another frequently
mentioned problem is turnover of the often small
manufacturing concerns in this field . In practice, this is of minor significance due to the continued entry of new vendors, usually with superior and/or
less expensive products. At present, there are at
least five or six reliable vendors of all of the more
critical system components (CPU, I/O boards, disk

controller boards, video interface boards). In most cases, ten or more vendors are available, and probably 30 or more companies make memory boards. The total number of boards available for the S-100 system is in the range of 200-300, far more than exist for any other type of computer. Accordingly, although several of the specific hardware items used in the present work are no longer marketed, functionally equivalent items are available in all cases.

The work described here was undertaken in the spring of 1976, only one year after the introduction of the first widely distributed personal computer (the Altair 880 sold by MITS, Inc.). In spite of the many advances made in the last four years (1976-1980), the only major changes we would suggest to someone building such a system would be use of the Z-80 (or possibly the new 16-bit 8086 or Z8000 chips) rather than the 8080 processor, and the use of a soft-sectored 8-inch floppy disk drive. The latter unit is compatible with the CP/M disk operating software system, which is powerful, convenient, moderately priced, and widely accepted as a de-facto standard by serious workers using S-100 computers. The CP/M system is configured in such a way that it is nearly independent of hardware variations within the many 8080 and Z-80 based microcomputer systems. (CP/M can also be used unmodified on the Exidy Sorcerer "appliance-type" microcomputer, on S-100 systems using 5 1/4-inch floppy disks with minor modifications, and on the widely distributed Radio Shack TRS-80 system in an extensively modified form.) Further discussion of the CP/M operating system can be found elsewhere.[13]

Finally, we would suggest consideration of a computer with no front panel, but with a ROM monitor program to provide access to memory and registers, etc. This allows a saving of several hundred dollars with no real loss in versatility.

Acknowledgement. Work of the University of Florida Radiation Chemistry laboratory is supported by the Department of Energy under Contract Number DE-AS05-ER03106. This is Document Number ORO-3106-72.

Summary. An inexpensive data acquisition and
reduction system for pulse radiolysis, flash photolysis on other transient chemical analysis has been
developed using mostly commercially available components. An optical signal from the kinetics experiment
is transduced via a photomultiplier tube, amplified
by a fast operational amplifier, and digitized at
rates up to 10^7 samples per second by a Biomation 610B
transient digitizer. The output of this device is
coupled by a simple logic interface into a parallel
port on an S-100 bus microcomputer. The configuration
of the IMSAI-8080 microcomputer includes 48K of solid
state memory, a floating-point mathematics board, a
series/parallel interface board, a high resolution
graphics system with keyboard input and video display,
an 8-inch floppy disk system, and a dot-matrix printer
capable of graphical output. The Biomation data
transfer routine is written in assembler, the graphics
package in PLM, and the data reduction package in
BASIC. Data can be accepted, displayed, fitted to
standard equations and plotted as desired, with print-
out of intercept, slope, and standard error data. The
entire system as described can be functionally duplicated for about $7,000.

References

1.   A general survey is given by H. Ruppel and H.T.
     Witt in "Methods in Enzymology, Vol. XVI. Fast
     Reactions," edited by K. Kustin, Academic Press,
     New York, 1969, pp. 317-379.

2.   A comparison of microsecond, nanosecond, and pico-
     second flash photolysis techniques is given by
     M.A. West in "Creation and Detection of the Excited
     State," vol. 5, Marcel Decker and Co., New York,
     1976, pp. 217-300.

3.   M. Ebert, J.P. Keene, A.J. Swallow, and J.H.
     Baxendale, editors, "Pulse Radiolysis," Academic
     Press, London and New York, 1965.

4. R.K. Wolff, M.J. Bronskill, J.E. Aldrich, and J. Hunt, J. Phys. Chem., <u>77</u>, 1350 (1973); J.E. Aldrich, P. Foldvary, J.W. Hunt, W.B. Taylor, and R.K. Wolff, Rev. Sci. Instr., <u>43</u>, 991 (1972); and earlier references to the work of Hunt's laboratory cited therein.

5. J. Lilie, J. Phys. Chem., <u>76</u>, 1487 (1972).

6. J.L.H. Patterson and S.P. Perone, Anal. Chem., <u>44</u>, 1978 (1972).

7. L.K. Patterson and J. Lilie, Int. J. Radiat. Phys. Cheml, <u>6</u>, 129 (1974).

8. T.E. Erickson, J. Lind, and T. Reitberger, Chemica Scripta, <u>10</u>, 5 (1976).

9. S. Gordon, K.H. Schmidt, and J.E. Martin, Rev. Sci. Instr., <u>45</u>, 552 (1974); K.H. Schmidt, S. Gordon, and W.A. Mulac, Rev. Sci. Instr., <u>47</u>, 356 (1976).

10. C.D. Jonah, Rev. Sci. Instr., <u>46</u>, 62 (1975).

11. A.T. Thornton and S.G. Laurence, Radiat. Phys. Chem., <u>11</u>, 311 (1978).

12. T. Barrett, D. Lunney, A.D. Salt, and M. Walter, J. Chem. Ed., <u>56</u>, 67 (1979).

13. A.R. Miller, Interface Age, <u>3</u>, No. 7, 156 (July 1978); <u>3</u>, No. 12, 130 (December 1978).

A Microcomputer System for Laboratory Automation

INTRODUCTION

Instrument control, data acquisition, data processing and report generation (final data analysis and display) constitute the primary task responsibilities of analytical laboratories. Usually these tasks are separately assigned and performed either manually or partly through utilization of facilities outside the laboratory (e.g. batch data-processing by a central host-computer). Alternatively, a dedicated laboratory minicomputer can be applied to automated control of several of these tasks. However, integration of a minicomputer to laboratory instrumentation usually requires considerable expertise in the design, construction and testing of the mechanical, electrical and logic interfaces to laboratory equipment. Recent advances in microcomputer systems have essentially eliminated many of the complexities associated with laboratory automation. This paper describes application of an interactive microcomputer system to automated operation of a quadrupole mass spectrometer that enables all the tasks outlined above to be accomplished under programmed control directly in the laboratory.

SYSTEM ELEMENTS

Mass Spectrometer

A schematic of the quadrupole mass spectrometer, used in this laboratory for investigation of polymer decomposition kinetics,[1] is shown in Figure 1. The polymer material is placed in a sample cell which is attached directly to the inlet flange of the spectrometer system. Thermal decomposition of the sample is performed under controlled heating rates either invacuo or in atmospheres of different composition. Volatile fragment species evolved during decomposition of the polymer chain structure are dynamically sampled through a small orifice by forming the fragments into

26

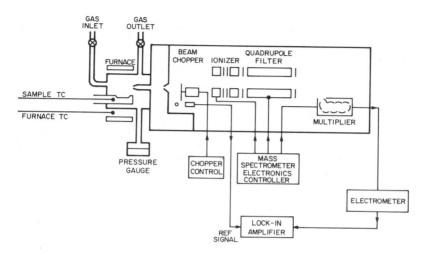

Figure 1   Experimental schematic of quadrupole mass spectrometer system used for thermal analysis measurements.

a molecular beam. The resulting sample beam is modulated by the mechanical chopper, ionized by electron impact, and mass analyzed. Beam modulation increases detection sensitivity and automatically eliminates interferences from background signals. In a typical pyrolysis experiment the sample temperature is programmed from 25-500°C at a rate of 5-10°C/min. Time- and temperature-resolved partial pressure profiles of the pyrolysis fragments are then obtained by repetitively scanning the mass spectrum throughout the entire course of sample pyrolysis.

Clearly, the first priority of the spectrometer/computer interface is to provide for automatic acquisition and analysis of the large amount of data associated with the accumulated mass spectra. Similarly, in the control mode the primary task is computer-control of the mass scan function. However, computer control and monitoring of peripheral mass spectrometric functions is also required (e.g. command of the chopper oscillator drive, gas inlet/outlet valves, pumps, etc.). Finally, multiplexing the input to the A/D converter used for the ion signal conversion enables acquisition of other analog voltage signals (e.g. temperature, pressure, etc.).

System Controller

A Tektronix Model 4051 desktop graphics system serves as the overall system controller. This is an M6800 microprocessor-based microcomputer containing 32K-bytes of random-access-memory (RAM), integral magnetic tape cartridge drive (300 K-byte storage capacity), and high-resolution CRT display (1024×780 addressable points). All real-time operations are directed by the 4051 via software commands to the data system, which in turn generates the necessary output signals for instrument control and data acquisition. The 4051 microcomputer is programmed in an extended BASIC language which provides compact graphics commands (e.g. DRAW, AXIS, ROTATE) and control of input/output (I/O) operations via PRINT and INPUT statements.

Use of a high-level interpretive language such as BASIC, and real-time display of alphanumeric and graphical output on the 4051 CRT screen, enables interactive control, through keyboard commands, of the experimental and data acquisition functions. In the control-mode the system operates on a priority-interrupt basis during which operation of the current program is halted, program and processor status stored, and control transferred to an appropriate interrupt-servicing subroutine. Upon completion of the service subroutine, the status of the original program is reloaded and operation continued from the point of interruption.

Input/output (I/O) operations are performed via two standard interface protocols. A bit-serial RS-222-C interface[2] enables communications with a central host-computer and I/O transfers to printer units. In addition, an IEEE-488 bit-parallel interface[3] (general purpose interface bus, GPIB) enables up to fifteen peripheral devices to be connected to the system under the command of a single controller. Each device on the GPIB is assigned a specific address, and on I/O transfers is designated by the system controller as a talker or listener. For example, the 4051 controller may command a digital voltmeter to transmit data to the bus (talker) and designate a printer to receive the data (listener). With the standardized IEEE-488 bus system various instruments (e.g., the data system, DVM's, power supplies, etc.) are interfaced to the 4051 microcomputer by simple cabling together the different components. GPIB cables and plug-compatible instrumentation are presently available from a number of different manufacturers.

Data Acquisition System

Digital conversion of the analog input signals is performed with a 16-channel microprocessor-based (8080A) data acquisition system equipped with 16 K-bytes of on-board memory. The system is composed of standard modular I/O boards, bus interface cards and memory boards[4,5] enabling

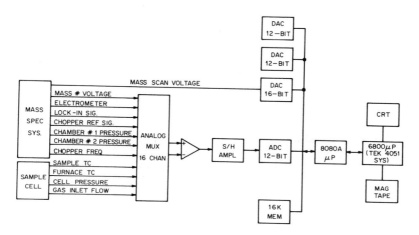

Figure 2   Block diagram of microprocessor-controlled data acquisition and instrument control system.

straightforward modification and system expansion. A list of the general specifications of the data system is presented in Table I, and a block diagram of the system is shown in Figure 2. The analog signals are scanned with the multiplexer (either sequentially or at random) and sampled with a 12-bit A/D converter. Autoranging of the signal multiplexer output from a X8 programmable gain amplifier provides an effective 15-bit dynamic measurement range (a separate low-level channel incorporating a X128 gain amplifier extends this to an effective 19-bit range).

The maximum data acquisition rate is 1000 Hz/channel. Two data transfer modes are possible: 1) direct data transfer to the 4051 microcomputer at low sampling rates (<100 Hz); 2) buffered data transfer of up to 16K-byte blocks of digitized samples using the data system's on-board memory (four separate 4K-byte RAM boards). This data buffer memory can be increased, if necessary, by simply replacing the 4K-byte RAM boards with 16K-byte boards. An increase in the direct data transfer rate to the 4051 is also possible by utilizing a recently developed hybrid assembly language[6] for the 4051 microcomputer to control transactions on the GPIB data bus. The 4051 microcomputer contains an integral magnetic tape cartridge drive that provides capability for permanent storage (300 K-byte capacity) and easy retrieval of the digitized data.

Real-time voltage control signals are generated from two 12-bit D/A

TABLE I

Data System Specifications

Number of Input Channels	32 single-ended or, 16 differential
Input Signal Range	0-10V
Programmable Gains	1,2,4,8 software selectable or autoranging
Sample Rate	0-1000 Hz
Acquisition Start	Manual or automatic
Data Resolution	12-bit (1 part in 4096)
Data Buffer Memory	16 K-bytes
Analog Output Channels	2, 12-bit DAC's 1, 16-bit DAC

converters and one 16-bit D/A converter included in the system. Program control of the mass scan rate and spectral range of the mass spectrometer, for example, is accomplished through resident EPROM (erasable, programmable, read-only memory) routines in the 8080 microprocessor, that produce an accurate output voltage ramp from the 16-bit D/A module. The ramp voltage end-points are selectable in the range 0-10V, with a minimum period between voltage steps of 1 msec. Other firmware routines enable threshold voltage selection, autoranging or fixed gain per channel, peak detection, and individually selectable channel scanning rates. With the latter, for example, channels 1 and 5 may be scanned at 500 Hz, channel 2 set inactive, channels 3 and 4 scanned at 100 Hz, etc. The firmware can be changed at any time in a similar manner as software is changed in other systems.

SYSTEM OPERATION

A schematic of the overall data acquisition and instrument control system is shown in Figure 3. The 8080-microprocessor data system, as well as other peripheral equipment (an X-Y plotter, waveform generator for con-

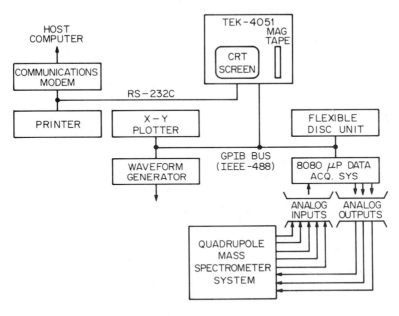

Figure 3   Overall schematic of the laboratory microcomputer system.

trol of the mechanical chopper, and external magnetic tape cartridge drive for data storage), are connected to the 4051 microcomputer via GPIB bus cables. An RS-232C line is used to interface with an impact printer for hard-copy output. Remote connection to a host-computer for off-line analysis of the digital data is also possible using the RS-232C interface.

In general, the 4051 controller is used to initialize all variable parameters (channel selection, sampling rate, etc.), store digitized data, perform final data manipulation and display the data. The 8080-based data system accepts and stores the initialization values, acquires the data, preprocesses the data (e.g. software scaling, threshold determination, peak detection, etc.), buffers the data, and transfers data to the 4051. The system is programmed to operate in an interactive mode with commands to the data system being issued in response to prompts and system status messages displayed on the 4051 CRT screen.

Computer-control of the mass spectrometer enables acquisition of the spectral data by several different techniques: 1) full scan of the entire spectrum; 2) partial scans of selected portions of the spectrum; 3) integer mass sampling. In each case the spectrometer is either tuned across a given mass range or to a specific mass by applying a voltage control signal (0-10V) to the quadruple mass filter rods. Mass tuning is approximately a linear function of the applied voltage. Ion transmission through the rod assembly occurs over the mass range 0-500 amu, resulting in a separation between adjacent mass number of approximately 20 mV. For precise mass control, therefore, a voltage resolution of better than 1.0 mV is required. This is achieved with the 16-bit D/A converter of the data system, which divides the voltage scan into 65,535 discrete increments of 0.15 mV (i.e. approximately 0.01 amu per voltage step).

In the full-scan mode the control voltage ramp parameters are input from the 4051 microcomputer in the form of initial and final mass numbers, voltage step-size, and ramp rate (in msec/voltage step). The digitized output from the analog signal is selected for either fixed gain or autoranging. Ion signal intensity data are then collected at each voltage increment of the D/A converter, unless the peak-detection mode is selected. In the latter case, the digitized ion signal data are pre-processed in real-time by a software routine resident in the 8080-microprocessor read-only memory prior to transfer to the system buffer memory. A fixed digital threshold may be set on the input signal in conjunction with this routine to decrease the appearance of spurious peaks due to baseline noise. Finally, the remaining analog input channels are initialized for channel gain and sampling rate, with the latter set to some multiple of the D/A ramp output increment (i.e. sampled at each voltage increment, or every other increment, or every third, etc.). This enables only as much data to be taken from an individual channel as is required. After all channels are initialized the acquisition command is sent and data automatically acquired from a pre-set number of voltage scans. At the end of each scan the data are transferred from the 8080 buffer memory to the 4051 magnetic tape unit.

A similar procedure is followed for operation in the partial-scan mode, with differences occurring only in the 4051 software control programs. Provisions are made to input initial and final mass values for several separate voltage ramps over the mass spectrum. In each individual ramp a limited number of mass peaks are sampled (usually from one to ten masses). Data collection in this manner is usually referred to as multiple-ion-detection(MID) or selected-ion-monitoring (SIM). since the mass selection parameters (e.g. number of mass groups selected, number of ions sampled per group) are set purely under software control, they may be easily changed to accommodate varying experimental requirements.

Integer mass sampling, in which the control voltage is incremented in one step from a given mass peak position directly to another, is also possi-

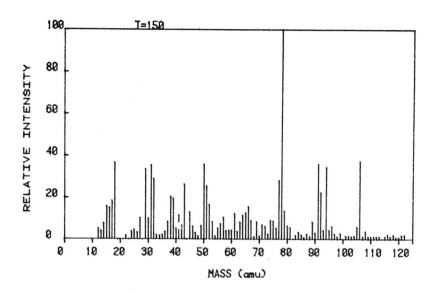

	m/e	I			m/e	I
(1)	78	100.0		(6)	31	35.4
(2)	106	37.5		(7)	94	34.7
(3)	18	36.7		(8)	29	33.7
(4)	91	36.2		(9)	32	29.1
(5)	50	35.9		(10)	77	28.4

Figure 4   Computer-generated plot of digitized mass spectrum obtained
from an epoxy polymer sample at 150°C. The peak intensity
data are normalized to the mass 78 signal, and a tabular listing
(optional) presented of the ten most intense peaks in the spectrum.

ble. This technique requires a calibration run to determine the exact
voltage-mass number relationship. With this technique improved peak
intensity S/N ratios are possible since the signal may be integrated for an
extended period at each mass. This is especially valuable in modulated
beam measurements for integration of weak signals detected with the lock-
in amplifier. A disadvantage of integer mass sampling, however, is drifting
of the peak voltages with time. In practice, this voltage drift is usually

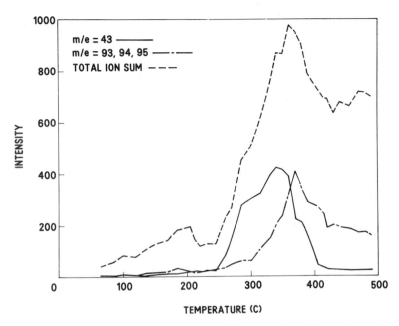

Figure 5   Computer-generated ion signal plot overlays illustrating selection
of profiles characteristic of a single ion (mass 43), selected ion
sum (masses 93, 94 and 95), and total ion current.

accommodated for by repetitively scanning the control voltage over a limited fractional mass range ($\sim$0.2 to 0.5 amu) on each side of the expected
peak center.

Once the spectral data are transferred to magnetic tape storage, powerful
digital processing techniques (e.g., data smoothing by digital filtering[7,8];
spectra subtraction and ratioing[9,10]) can be implemented. Final data reports
are generated in the form of tabular listings of mass and peak intensity
values, bar-graph spectrum plots of absolute or normalized peak intensities
(the digitized data may be normalized to the most intense peak of a spectrum, or to any peak in the spectrum), selected ion profile plots as a function of time or temperature (any number of ion profiles may be overlayed
on a single plot), and three-dimensional isometric plots of the ion signal
intensity as a function of both temperature and mass number.

Examples of these data output options, shown in Figures 4-7, were generated directly from the system X-Y plotter under control of an interactive

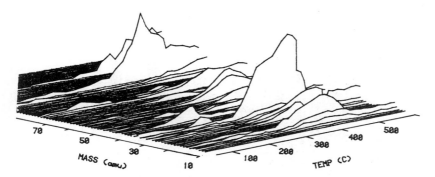

Figure 6   Three-dimensional isometric plot of ion signal intensities as a function of both mass number and temperature.

display program run on the 4051 microcomputer. A normalized mass spectrum plot is shown in Figure 4 in which an optional listing of the ten most intense peaks in the spectrum was also selected. Figure 5 presents ion profile plots of a single ion (mass 43), selected ion sum (masses 93, 94 and 95), and total ion current. These plots were generated by computer-search of the spectra for this particular pyrolysis run, stored on separate magnetic tape files, to retrieve the peak amplitudes at the specified mass values. When sample degradation is relatively extensive, a large number of volatile fragments are produced resulting in a fairly complex spectrum, such as in Figure 4, where signal peaks are observed at virtually every mass number. In these cases a three-dimensional isometric plot of the composite product yields is an important diagnostic aid as shown in Figure 6, since the overall thermal behavior can be discerned at a glance. This information can then be used to select specific individual ion profiles for more detailed analysis. Finally, a related isometric plot of the ion signal data is shown in Figure 7, in which the mass and temperature axes have been switched, as an illustration of the ease with which the ion data may be manipulated once in digital form.

SUMMARY

Microcomputer-based data acquisition and instrument control techniques have significantly enhanced the capabilities of a quadrupole mass spectrometer system. All processes from acquiring and storing data, as well as real-time control of the experimental apparatus, to final data processing and display are now performed directly in the laboratory. Data acquisition capabilities have been increased enabling more analog data to be collected at

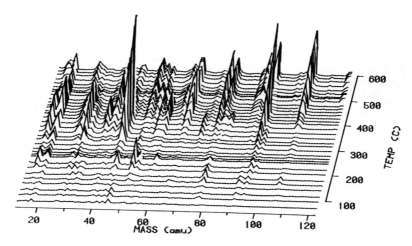

Figure 7   Isometric plot of ion peak intensities with mass and temperature
axes switched with respect to Figure 6.

higher rates and with greater accuracy than previously possible. Analyses,
which were otherwise unfeasible, are now performed in a routine manner.
System modification and expansion are easily accommodated due to the
modular design approach characteristic of both the hardware and software.

REFERENCES

[1]   R. M. Lum, J. Polym. Sci. Chem. Ed., *17*, 203 (1979).

[2]   EIA Standard RS-232-C, "Interface Between Data Terminal Equipment and Data Communication Equipment Employing Serial Binary
Data Interchange", Electronic Industries Association, 1969.

[3]   IEEE Standard 488-1975, "Digital Interface for Programmable Instrumentation", The Institute of Electrical and Electronic Engineers, Inc.,
1975.

[4]   Control Logic, Inc., Natick, Massachusetts.

[5]   US DATA Engineering, Dallas, Texas.

[6]   C. Rose, J. A. Copeland and L. G. Cohen, private communication.

[7]   A. Savitzky and M. J. E. Golay, Anal. Chem., *36,* 1627 (1964).

[8]   D. Binkley and R. Dessy, J. Chem. Ed., *56,* 148 (1979).

[9]   Y. S. Chang, J. H. Shaw, W. M. Uselman and J. G. Calvert, Appl. Opt., *16,* 2116 (1977).

[10]   C. L. Lin, W. M. Gutman, J. H. Shaw, W. M. Uselman and J. G. Calvert, Appl. Opt., *17,* 993 (1978).

W. A. FELD, C. R. SHORE,
and M. D. PORTER

CHAPTER 4

The Use of the Kim-I Microcomputer in an Automatic Titration Apparatus

Abstract. The low cost and ready availability of a microcomputer such as the KIM makes it attractive as a controller and data collection device in the laboratory. A simple automatic apparatus is described which is capable of performing potentiometric titrations in an incremental volume addition mode with volume addition leveled by proximity of the titration to the equivalence point. The apparatus is assembled from common laboratory equipment including an Orion 801A Ionalyzer with BCD input/output and in inexpensive Houston Glass automatic buret. The memory requirements are minimal, and all timing is done in software to minimize the interfacing hardware. The specific design, software and overall characteristics of the system are described with respect to various common titrations with emphasis on the set-up, data presentation and interpretation of the results. A comparison of the systems emphasizing available features, expandability and ease of use is given.

INTRODUCTION. The application of microcomputers in
the automation of potentiometric titrations has increased markedly in recent years. Several systems
have been developed utilizing different titrant delivery systems, types of control-hardware or software
and endpoint determination methods.
    Methods of titrant delivery range from a pulsed
droplet device (1) to gravimetric systems (2-5).
Each 'buret' has advantages and drawbacks. The
pulsed droplet device is inexpensive and time consuming. Gravimetric devices are efficient but possibly
not economical for some applications.
    Hardware controlled titrators are now commer-
cially available and also can be constructed in the
laboratory (1-5). Hardware systems in general, lack
versatility and require expertise in electrical design.
    Software or microcomputer controlled titrators,
have been developed and vary in complexity, versatility and cost. Programming has been implemented in
the assembly language instruction set of the microcomputer (6) or in a higher level language (2,7).
Such systems offer versatility and are now becoming
cost-competitive with hardware designs.
    Most endpoint calculations employ either the
Gran type method (3) or first or second derivative
techniques (2,7). Each method has advantages and
complexities in terms of programming. Typical applications include simple acid-base titrations (1-5,7),
ion selective electrode studies (5,8), and solution
equilibria investigations (6).
    This article describes a microcomputer-control-
led automatic titrator which was designed to perform
acidity titrations. The system is easy to construct,
i.e. it requires minimal digital logic expertise. It
is versatile in that a change in programming would
allow a wide range of acid and base titrations to be
performed. It is, in addition, an economically feasible device.

Buret System. The titrant delivery system, Figure 1,
is a slightly modified Electromatic Buret (No. 180000,
Houston Glass Fabricating Cl., Houston, Texas, 77011).
The buret consists of a glass encapsulated iron core
which is opened by means of a 115 VAC solenoid. The
solenoid is controlled by an optically-isolated relay

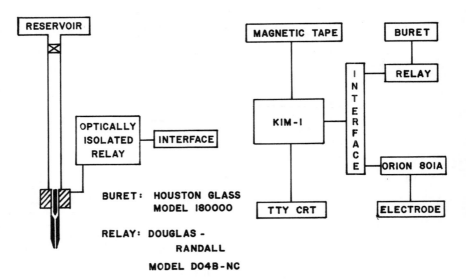

RESERVOIR

OPTICALLY ISOLATED RELAY — INTERFACE

BURET: HOUSTON GLASS MODEL 180000

RELAY: DOUGLAS - RANDALL MODEL DO4B-NC

FIGURE I. BURET SYSTEM

MAGNETIC TAPE

KIM-1

TTY CRT

INTERFACE

BURET

RELAY

ORION 801A

ELECTRODE

FIGURE 2. HARDWARE BLOCK DIAGRAM

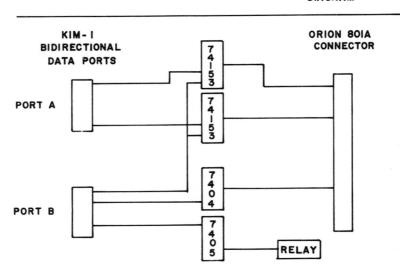

KIM-1 BIDIRECTIONAL DATA PORTS

ORION 801A CONNECTOR

PORT A

74153

74153

PORT B

7404

7405

RELAY

FIGURE 3. INTERFACE DIAGRAM

40

(Douglas-Randall, No. DO4B-NC). The relay allows a
digital system to control 115 VAC without the danger
of high voltage leaking into the microcomputer system.
To insure 'constant addition', the buret was attached
to a 20 liter titrant storage tank which was positioned above the buret to insure a constant pressure
head. The precision of the delivery system was determined gravimetrically and found to be about 1% for
a volume addition of 0.134 mls.

Orion 801A Meter. The progress of the titration was
followed using an Orion 801A pH/millivolt meter. The
meter can measure pH from 0.001 to 13.999 in increments of 0.001 units. Interfacing is available via a
standard 22/44 pin connector and is TTL compatible.
Meter output is given in BCD and also allows input
lines for several functions such as holding the ADC
for reading the meter. The electrode system was a
Laboratory Grade Orion Combination pH Electrode (No.
H5820-4). The combination electrode was chosen for
its ease in handling and compactness.

The KIM-I Microcomputer. The KIM-I (9) is a compact
single board microcomputer that employs the 6502
microprocessor and features 1K of RAM, 2K of ROM containing the system executive, a complete audio cassette interface, a serial terminal interface, 15 bidirectional I/O lines, a 23 key keypad and a six
digit LED display. Auxiliary equipment included a
power supply, a cassette recorder and a model 33 KSR
teletype. Memory expansion was accomplished by means
of a custom built S-100 bus interface (10) and two
8K S-100 RAM boards (11).

The KIM-I/Orion 801A Interface. The KIM-I/Orion
801A interface (Figure 3) consists of four parts: 1)
the KIM connection, 2) a data selection/buffering
section, 3) the Orion 801A connection and 4) buret
control.
       The 15 bidirectional I/O lines provided by the
KIM are arranged as two ports (A & B). Each line can
be separately programmed for input or output by writing a status word into the correct memory location.
We defined Port A for the input of data and Port B
for the output of control signals.

The Orion 801A data connector provides each
digit in a 4-line, BCD format and a "HOLD" control
signal which can be used to "freeze" the meter display. A "HOLD" signal was provided by one bit of
Port B buffered by two sections of a 7404 hex inverter.
The four digits were read by employing two dual
1 of 4 data selectors (74153), the outputs of which
were connected to the least significant four bits of
Port A. The most significant four bits of Port A
were set to a logic low. Digit selection was accomplished by using two bits from Port B as a two bit
binary counter.
Buret control (Figure 1,3) was accomplished by
toggling the least significant bit of Port B. This
control signal was buffered with one gate of a 7405
open collector inverter. An open collector device
was needed to provide the voltage swing necessary to
operate an optically isolated relay.

Software. All programming was done in the computer
language TINY BASIC. TINY BASIC is an abbreviated
form of the BASIC computer language available on
large scale computer systems and has a USR function
which provides a link to machine language subroutines
or, in our case, access to control and data-acquisi-
tion memory locations.
Once the titration process is initiated, the
microcomputer has three primary functions: 1) to control the delivery of titrant, 2) to read a stable pH
value and 3) to locate the inflection point of the
titration from the data generated. A flowchart of
the software is shown in Figure 4 and a listing is
shown in Figure 5.
Initialization of the I/O ports (Figure 5, line
6) defines Port B as output. On reset the ports are
initialized as inputs so no initialization of Port A
is necessary.
User defined inputs (Stability Loop Constant,
Timing Loop Constant and Volume Increment) are requested, stored, and reprinted for future reference
(Figure 5, lines 17-29).
Titrant delivery volume is controlled by the
BURET OPEN subroutine (Figure 5, lines 399-410).
BURET OPEN is a simple timing loop which opens the
buret, decrements the Timing Loop constant to zero

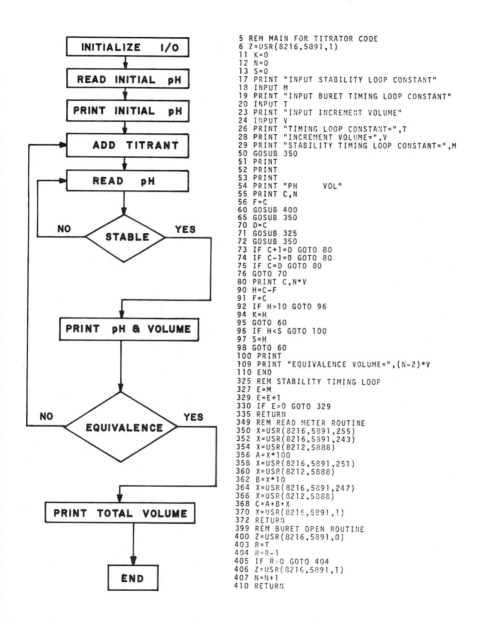

```
5 REM MAIN FOR TITRATOR CODE
6 Z=USR(8216,5891,1)
11 K=0
12 N=0
13 S=0
17 PRINT "INPUT STABILITY LOOP CONSTANT"
18 INPUT M
19 PRINT "INPUT BURET TIMING LOOP CONSTANT"
20 INPUT T
23 PRINT "INPUT INCREMENT VOLUME"
24 INPUT V
26 PRINT "TIMING LOOP CONSTANT=",T
28 PRINT "INCREMENT VOLUME=",V
29 PRINT "STABILITY TIMING LOOP CONSTANT=",M
50 GOSUB 350
51 PRINT
52 PRINT
53 PRINT
54 PRINT "PH        VOL"
55 PRINT C,N
56 F=C
60 GOSUB 400
65 GOSUB 350
70 D=C
71 GOSUB 325
72 GOSUB 350
73 IF C+1=D GOTO 80
74 IF C-1=D GOTO 80
75 IF C=D GOTO 80
76 GOTO 70
80 PRINT C,N*V
90 H=C-F
91 F=C
92 IF H>10 GOTO 96
94 K=H
95 GOTO 60
96 IF H<S GOTO 100
97 S=H
98 GOTO 60
100 PRINT
109 PRINT "EQUIVALENCE VOLUME=",(N-2)*V
110 END
325 REM STABILITY TIMING LOOP
327 E=M
329 E=E+1
330 IF E>0 GOTO 329
335 RETURN
349 REM READ METER ROUTINE
350 X=USR(8216,5891,255)
352 X=USR(8216,5891,243)
354 X=USR(8212,5888)
356 A=X*100
358 X=USR(8216,5891,251)
360 X=USR(8212,5888)
362 B=X*10
364 X=USR(8216,5891,247)
366 X=USR(8212,5888)
368 C=A+B+X
370 X=USR(8216,5891,1)
372 RETURN
399 REM BURET OPEN ROUTINE
400 Z=USR(8216,5891,0)
403 R=T
404 R=R-1
405 IF R>0 GOTO 404
406 Z=USR(8216,5891,1)
407 N=N+1
410 RETURN
```

FIGURE 4. SOFTWARE FLOW CHART   FIGURE 5. SOFTWARE LISTING

43

and then closes the buret. The addition volume is set by the magnitude of the Timing Loop constant.

The program uses the subroutine stability Timing Loop (Figure 5, lines 325-335) to determine when mixing, reaction equilibration and electrode response are complete. The program cycles until a pH value of + 0.01 pH units from the previous reading is obtained. The time between pH readings is controlled by the Stability Timing Loop constant.

The Orion 801A meter is read by a subroutine (Figure 5, lines 349-372) which makes use of the USR function in TINY BASIC. Values can be output to or read from a memory location as required. Because TINY BASIC cannot handle decimal points, no attempt was made to introduce them and all numbers appear as integers. Location 8216 (base 10) is the beginning of a subroutine used to place a number in a specified location. Location 5888 (base 10) is the data register for Port A and location 5819 (base 10) is the data direction register for Port B.

The endpoint of the titration is determined by calculating the first derivative of the titration data and finding the maximum value for the function (Figure 5, lines 90-98). To circumvent noise problems, a minimum pH rise must be observed (Figure 5, line 92) before the derivative maximum is determined.

The titration ceases after the first derivative maximum is passed by several additions. The equivalence volume is then printed and control is returned to BASIC. A typical printout is shown in Figure 6.

RESULTS. Titrations using the described system were performed on hydrochloric and acetic acid with sodium hydroxide as the titrant. The results are given in Figure 7 and represent an average of five determinations. The accuracy is easily in the acceptable range.

CONCLUSIONS. Times for each titration are also shown in Figure 7. While the system is not fast, a few changes could improve the simple analysis rate. Improvements in the system can be made easily by more efficient programming. The software could be modified so that large volumes of titrant are added initially with smaller volume increments used as the endpoint is approached. This, of course, implies

```
:RUN
INPUT STABILITY LOOP CONSTANT
?100
INPUT BURET TIMING LOOP CONSTANT
?1
INPUT INCREMENT VOLUME
?134
TIMING LOOP CONSTANT=   1
INCREMENT VOLUME=       134
STABILITY TIMING LOOP CONSTANT=  100
```

PH	VOL
600	0
602	134
603	263
601	402
602	536
608	670
609	804
610	938
612	1072
613	1206
610	1340
611	1474
612	1608
618	1742
619	1876
620	2010
622	2144
623	2278
620	2412
621	2546
622	2680
623	2814
629	2948
629	3082
631	3216
632	3350
631	3484
633	3618
630	3752
632	3886
633	4020
639	4154
640	4288
642	4422
643	4556
641	4690
643	4824
649	4958
652	5092
651	5226
658	5360
660	5494
672	5628
682	5762
732	5896
760	6080

```
EQUIVALENCE VOLUME=        5762

:
```

FIGURE 6. SAMPLE PRINTOUT

TITRATION OF HCL (0.01000 N)

RUN	VOLUME (ML)	
1	6.164	
2	6.030	
3	5.896	
4	6.030	TIME = 8 MIN.
5	6.030	2% ERROR

AVERAGE N = 0.01022

TITRATION OF HAc (0.01920 N)

RUN	VOLUME (ML)	
1	4.556	
2	4.422	
3	4.556	
4	4.556	TIME = 6 MIN.
5	4.690	0.5% ERROR

AVERAGE N = 0.01931

FIGURE 7.   TITRATION   DATA

COMPUTER HARDWARE

KIM-I	$170
POWER SUPPLY	50
MEMORY (8K)	120
S-100 BUS	30
CASSETTE	80
TTY	600
	$1050

PERIPHERAL SYSTEM

BURET	$125
METER	1200
INTERFACE	15
ELECTRODE	45
	$1385
SYSTEM TOTAL	$2435

FIGURE 8.   SYSTEM   COST

some previous knowledge of the sample or an elaborate computational routine. The titrant concentration could be changed, but is time consuming.

   User interaction time could be reduced by incorporating an automatic sample handling device.

   The accuracy of the endpoint determination could also be improved by a software adjustment. At present, the equivalence volume is taken as the volume added up to the calculated peak maximum jump. An interpolation routine would be expected to improve the accuracy of the titration.

SYSTEM COST. Figure 8 gives approximate costs of the components incorporated in the system. The prices may vary, but with selective purchasing these figures can be met. One possibility for cost improvement would be to use an ADC card to replace the pH meter. This would drop the peripheral costs to about $500. Also it is preferable to have a CRT terminal available as this greatly shortens the time needed to write and debug the software. The KIM microcomputer can operate at terminal speeds up to 9600 Baud.

REFERENCES.

(1) Hunter, T.W., I.T. Sinnamon, and G.M. Hieftje, Anal. Chem., 47, 497 (1975).

(2) Martin, C.R. and H. Freiser, Anal. Chem., 51, 503 (1979).

(3) Johansson, A. and L. Pehrsson, Analyst, 95, 652, (1970).

(4) Mitchell, D.G. and K.M. Aldous, Analyst, 98, 580 (1973).

(5) Christiansen, T.F., J.E. Busch, and S.C. Krogh, Anal. Chem., 48, 1051 (1976).

(6) Leggett, G., Anal. Chem., 50, 718 (1978).

(7) Wu, A.H.B., and H.V. Malmstadt, Anal. Chem., 56, 2090 (1978).

(8) Gaarenstroon, P.D., J.C. English, S.P. Perone, and J.W. Bixler, Anal. Chem. <u>50</u>, 811 (1978).

(9) The KIM microcomputer is produced by Commadore Business Systems, Inc., 950 Rittenhouse Rd., Norristown, PA, 19401.

(10) Eaton, J., <u>Kilobaud</u>, <u>1</u>, 36 (1978).

(11) Econoram II boards are available from Godbout Electronics, Box 2355, Oakland, CA, 94614. Other S-100 RAM boards may be substituted.

E. T. GRAY, JR., and H. J. WORKMAN   CHAPTER 5

Fundamentals of a Z-80-Based
Electrochemical System

Introduction

The instrumentation for voltammetry and coulometry requires
only three electronic components, a voltage source, a current
to voltage converter, and a voltage amplifier. Virtually all
forms of voltammetry and coulometry would be available from one
instrument 1) if it had a voltage source from which any voltage
as a function of time could be obtained, 2) if it could be
operated chronopotentiometrically or chronoamperometrically,
3) if any function of the applied voltage and measured current
could be displayed graphically, and 4) if options varying from
totally automated to totally manual control were possible.

This versatility cannot be provided by a "classical" micro-
processor-controller but can be provided by a microcomputer.
Some microcomputer manufacturers have the necessary hardware
and software available which hitherto was available only with the
more expensive minicomputers. From the available microcomputers,
the TRS-80 was chosen for a number of reasons: 1) The interfacing hardware required is available and inexpensive; the common support chips designed for the 8080 processor are compatible
with the Z-80 chip (the heart of the TRS-80 system). 2) The
instruction set of the Z-80 microprocessor is well designed for
simple interfacing. 3) A 40-pin-out edge-card is available to
which peripherals can be directly connected. Interfacing is especially easy when the external devices are mapped as ports. 4)
Although BASIC is provided with the system, FORTRAN and an
Editor/Assembler are available as options. With all three
languages available a full range of software development is
possible.

48

In order to develop the software and hardware needed for an electrochemical unit efficiently,   it is necessary to have a complete microcomputer system or what may be called a "development system." For the TRS-80 this would include the Z-80 microprocessor, level II and DISK-BASIC, 12K ROM, 48K RAM, expansion interface, 5" floppy-disk, video display, and a line printer.

A whole development system (~$3400) need not be dedicated to an electrochemical station, however. As little as the Z-80 microprocessor (video display and tape cassette), level II BASIC, 12K ROM and 16K RAM (~$850) is sufficient to run the unit but data storage on disk and line printer output are sacrificed along with some minor advantages which DISK BASIC offers. This minimum system would be painfully slow for software and hardware development work and FORTRAN programs cannot be developed without a disk. It is important to realize that the "development system" has the capabilities normally associated with minicomputer architecture and it is this very point we have exploited in our design.

Fundamentals

The basic philosophy in the design of the apparatus is that the computer would be in full control. The only options not directly controlled by the computer are 1) a switch to select a 2-electrode or a 3-electrode system, 2) a switch to select an anodic or cathodic scan, 3) a switch to engage a standard resistor in place of the cell, 4) a switch to select high or low sensitivity, and 5) a switch for main power.

The main sections of the unit (as shown in Figure 1) are a digital-to-analog converter (D/A) system to generate the applied voltage, a current-to-voltage converter, whose offset is directly controlled by the CPU through a D/A, an amplification and filtering cascade with resistors and capacitors that can be engaged or disengaged by the CPU through CMOS switches, an analog-to-digital converter (A/D) board to bring faradaic information into the computer, and a variety of output options including the standard X-Y recorder, video display (48 x 128 resolution) and line printer. Each of these sections will be discussed individually.

Voltage Applied. The applied voltage is generated from a D/A whose analog signal is modified by a series of four operational amplifiers. The first decision to be made in designing this

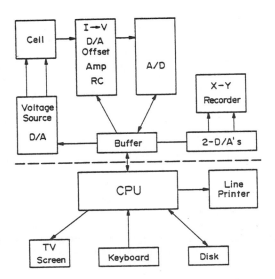

Figure 1. Block diagram of the electrochemical unit. The TRS-80 microcomputer system is below the dashed line, where CPU is the combination of the keyboard and the expansion interface. The 40-pin bus of the CPU is put through a bidirectional buffer in order not to overload the system.

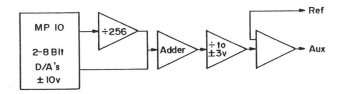

Voltage Source

Figure 2. Electrochemical voltage source. One of the two 8-bit D/A's is attenuated by 256 and added to the other channel to give the equivalent 16-bit precision. All the amplifiers are common operational amplifiers, the last stage being used as a voltage follower. Only 17 computer lines (A0-A7, D0-D7 and OUT) are needed to control the MP10, and these are connected directly through the buffer.

component is the choice of D/A's. Most voltammetry is limited
to the range from +3 to -3 volts by the electrode materials
and solvents. If it can be assumed that 300 mv is the narrowest
window ever to be viewed out of the 6000 mv available, it is
reasonable to require a D/A with a 1 mv minimum step size. This
is 1 part in 6000. Since 2^{12} < 6000 < 2^{13}, at least a 13-bit
D/A is needed.

   In order to accommodate this, a Burr-Brown microperipheral
MP10 was chosen. It is a dual 8-bit D/A with a ±10-volt
range. As can be seen in Figure 2, one D/A channel is attenuated by a factor of 256, inverted, and then added to the other
channel which is unattenuated, effectively giving a 16-bit D/A.
Although this may seem like a tortuous route, the MP10 contains
all the necessary hardware to allow it to be directly linked
to the Z-80 bus without any other hardware. (Those who are very
fond of wire wrap mazes may wish to choose another route as is
indeed done in other sections of this system.)

   The output of the D/A board is an analog voltage determined
by the combination of the most significant bits (MSB) (voltage unattenuated) and the least significant bits (LSB) (voltage attenuated by 256). Zero volts is obtained by setting
the MSB to 127 (7FH) and LSB to 0. After the addition of the
two voltages the resultant is then dropped into the ±3 volt
range. Therefore the range of the MSB's is set at ±3 volts
which defines the working range as ±(3 + 3/256) or ±3.0117 volts
in steps of 1 part in 2^{16}.

   A voltage step of 1 part in 2^{16} requires comment
in terms of the noise of the system. The output of the D/A normally has a ±1 mv ripple. The LM324 op amps normally would not
add to this significantly in the configurations shown in Fig. 2
so long as the power and ground leads have capacitance across
them at each chip. Unfortunately, 25 mv of noise actually ap-
peared at the output of each amplifier stage, irrespective of
input to that stage. This was found to be entirely "computer
noise," not only from the microprocessor itself, but also the
buffer chips, address decoders, ports, and every other device
that can respond to the constantly fluctuating address lines
from the Z-80 bus. This noise from digital sources can be
pared down to about 5 mv by judicious use of RC filtering.
Also, by halting the CPU during a track-and-hold and subsequent A/D conversion, the noise can be reduced so that at least
12 significant bits can be realized. Moreover, most electrochemical techniques will successfully measure the average current generated by the ripple so long as the ripple is not more
than a few millivolts, thus increasing the effective number of
significant bits.

After attenuation, the signal is passed to a voltage follower where the feedback connection is the reference electrode, and
the output of the op amp is the counter electrode. These
leads are connected via a manual switch in a 2-electrode configuration.

Current-to-Voltage Converter. The current-to-voltage converter
is the first stage of the cascade in Figure 3. Either a
$50K\Omega$ or a $1K\Omega$ resistor may be engaged for initial amplifications
of 5×10^4 or 10^3, respectively. Since the A/D to be used is
in a 0-5 volt configuration, the final output of this stage
must be in this range.

A current offset is applied to this amplifier so that a small
signal on a big background can be offset before amplification.
This is accomplished by a 12-bit D/A in the ±5 volt configuration   (a resolution of ~2.5 mv/bit). To control the D/A requires 12 of the 24 (I/O) lines of an 8255 programmable
parallel-interface-port. Therefore, as with the MP10, the 8255
must be addressed twice for a complete update of the signal.
The DAC-80 plus the hardware necessary to address it are about
$45 (in comparison to $100 for the MP10).

Amplification and Analog Filtering. The amplifier cascade consists of simple inverting amplifiers which have variable resistive
elements in the feedback loops so that the amplification can be
changed. These resistors are engaged or disengaged
by means of CMOS switches. One end of each switch is connected
to the virtual ground at the "-" input of the amplifier stage.
The other end is connected to the feedback resistor. These
switches do not function when one end is connected to the output of the op amp instead of the summing point. Also, the voltage necessary to power the switch must be greater than the voltage to be switched. Therefore, an op amp is used to provide
$7\frac{1}{2}$ volts to power the switches. Unfortunately, at this voltage,
the 3.7-4.0 volt high from the CPU and support chips is not
sufficient to control the switches.Consequently,the high needed
to turn the electronic switch on must be raised to 5.0 v. This
is done with SW7417 chips.

Each switch is controlled by one of the remaining 12 (I/O)
lines of the 8255 (recall that the other 12 lines control the
offset D/A). Thus the computer can fully control offset and
amplification. Therefore a well behaved system could be analyzed automatically once the sample is prepared, the elec-
trodes introduced into the solution and the voltage range (and
direction) selected.

0.1 msec time constant is built into the circuitry but a
0.1 sec and 1 sec time constants are selectable through switches
11 and 12. These switches are placed at the virtual ground as
in the case of the resistors.
<u>Analog to Digital Converter</u>. Two alternatives are offered in
Figure 4 and both have been used at different times. The MP 22
(A/D) is another Burr-Brown microperipheral. This 80-pin chip
does not have a sample-and-hold but it can be added if desired.
Otherwise only resistors and capacitors must be added to complete this 12-bit A/D. The device has a 16-channel input
multiplexor. This is somewhat of an extravagance if the A/D
is dedicated to the electrochemical unit, unless other measurements are to be made simultaneously such as temperature or pH.
On the other hand, a single channel A/D such as the ADC-80
pictured in Figure 4, requires the usual address decoder, 8255
port, and assorted one-shots to trigger conversion. Note that
the programmable clock shown in Figure 4 makes the system less
software-bound as the interrupt timing is taken care of by the
hardware. However, a 25 msec interrupt from the Expansion
Interface is available and could be used for timing purposes.
Software counters could make use of this interrupt to determine
when data should be taken.
The MP22 can be quickly interfaced to the Z-80 because it
is a complete package. However, hardware modifications are
not easy or even possible in many applications. Therefore,
software needed for control can become cumbersome in some
situations. In the ADC-80 only the converter is provided, but
interfacing can be accomplished by using appropriate address
decode, etc. Since the hardware is configured as desired by
the user, software control of the A/D can be simplified.
<u>Output</u>. Two forms of output are part of the microcomputer sys-
tem itself. The video display, although low in resolution,
does provide a way of viewing the data so **as to** assist in making a decision about data quality. A negative decision could
mean simply clearing memory to begin again. A favorable decision then leads to further decisions regarding analysis of the
data by the computer, reproduction of the curve <u>via</u> X-Y recorder output or strip chart output, or some form of line printer
presentation of the numerical data and/or conditions. The two
12-bit D/A's (DAC-80) used to generate the analog signals for
X-Y recorder output are interfaced in the same way as the D/A
used as the current offset in Figure 3. A voltage follower is
used to protect the D/A's from overload. If a pen-up/pen-

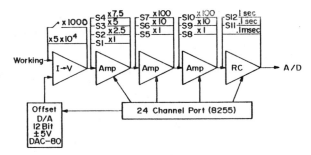

Figure 3. Current measuring cascade. All amplifiers are common operational amplifiers. Unity gain amplifiers which are
placed after each amplifier for offset purposes are omitted
from the figure for clarity. The 8255 multiplexes 24 data lines
are 3 ports of the Z-80 (the 8255 is addressed by a 74LS138
address decoder not shown). Twelve of these lines are data
lines for the DAC-80 which supplies a ±5V offset. The other
twelve lines operate 12 CMOS switches (CD4066) which activate
the amplification and time constant feedback circuits. A unity-
gain-amplifier after the current follower can manually be
engaged as a polarity switch.

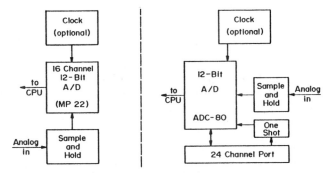

Figure 4. Two potential analog-to-digital configurations.
1) Using the MP22 12-bit microperipheral, eight address lines,
eight data lines, an IN line and the WAIT line are connected to
the CPU buffer. The few resistors and capacitors necessary are
indicated by the manufacturer. 2) The ACD-80 is connected to 12
of the 24 lines of the 8255. One more line is used for the
one-shot to initiate conversion. As in Figure 3, the 8255 is
controlled by a 74LS138 address decoder (not shown).

54

down accessory is available, one line from the 8255 can be used
to control the pen. This leaves 11 lines for the 12-bit D/A
that generates the analog signal for the X axis. Therefore,
the resolution for the axis is 1 part in 2000.

Results and Discussions

Figures 5 and 6 show the response of the system under various
conditions but at minimum amplification. This experiment was
performed before the 1KΩ resistor was introduced as an option
at the current-to-voltage converter. Figure 5 illustrates
polarograms of 5 x 10^{-4} $\underline{M}$ Cu(II) in 1.0 $\underline{M}$ NH$_3$. The voltage
signals were monitored by an oscilloscope before they were fed
into the A/D. In both cases, the voltage staircase applied to
the cell had 70 equal steps. In Figure 5a, a two-electrode
system was used, the counter electrode being a mercury pool.
Figure 5b was generated on the same solution as 5a but a
saturated calomel electrode was added to the cell to create a
3-electrode system.

For these polarograms, 5 volts at the A/D correspond to 100μ
amps. Therefore, the wave in Figure 5a is off-scale so far as
the A/D is concerned. The 1KΩ resistor was inserted to eliminate
this problem. This resistor was inserted with a manual switch
since the CMOS switch caused difficulties in the absence of the
virtual ground.

Figure 6 illustrates a pulsed-polarogram using a mercury
drop as the working electrode and a mercury pool as the counter
electrode. As in Figure 5, the amplification factor of 5 x 10^4
is used and 5 volts at the A/D correspond to 100μ amps. Each
pulse increased V_{app} by 10 mv and returned to a resting voltage
of zero volts.

In most analog pulsed-polarographs, the time constant on the
sample-and-hold corresponds to roughly five pulses. The data in
Figure 6 were not manipulated in this manner. This is a display
of the data which the computer received from the A/D. One of
the advantages of the digital system is the ability to choose
the width of the window in the moving-window averaging technique
which could be applied to the digital data to mimic the time
constant used in the analog system. Since the data are stored,
this choice can be made after the experiment is over.

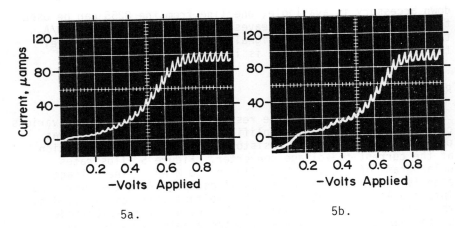

5a.   5b.

Figure 5. Polarograms of 5 x 10^{-4} $\underline{M}$ Cu(II) in 1.0 $\underline{M}$ NH$_3$ using
Triton-X as maximum suppressor. Both polarograms were generated
with a staircase of 70 steps, RC = .1 sec and a drop life of
~2 sec. Figure 5a. shows the results of a 2-electrode system
(Hg drop $\underline{vs}$. Hg pool), and 5b. a 3-electrode system (Hg drop
working, $\overline{Hg}$ pool counter and saturated calomel reference).

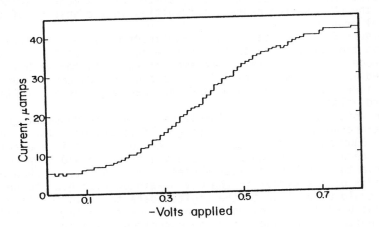

Figure 6. A pulsed-polarogram of 30$\mu\underline{M}$ Cu(II) in 0.1 $\underline{M}$ NH$_3$ with
a resting voltage of zero volts and a voltage increase of 10 mV
per step (Hg-drop $\underline{vs}$. Hg pool). The time constant is 0.1 sec.
In order that the $\overline{data}$ could be viewed directly, no type of
signal averaging was used (as is normally done in commercial
instruments).

Conclusions

The list of advantages of an electrochemical unit whose basis is a microcomputer system is lengthy. Since the CPU generates the applied waveform, changing from one form of voltammetry to to another is just a matter of changing part of the software. Also controlled by the computer are the current/voltage offset, voltage amplification and filtering (both analog and digital). There is sufficient control to allow a complete voltammogram to be obtained with minimal operator interaction. Simply having the data transferred to computer memory means the data can be analyzed rapidly and/or stored on disk. Along with the classical X-Y recorder output, video display and line printer copy are available. The whole system is very inexpensive. The electronics run from $400 to $600 and the computer system from $850-$3400.

Finally, the system can be rapidly constructed. Since the computer requires no construction or modification, only the construction of the appropriate electronic boards is required. Our system was designed and constructed in three months.

C. R. BEAVERS, S. E. GEORGE,
J. L. ROBINSON, and J. R. WRIGHT

CHAPTER 6

Linking Several Moderate Usage Laboratory Instruments to One Microcomputer

INTRODUCTION

In 1977, plans for purchasing a signal averager for our NMR Spectrometer led first to a consideration of commercial equipment designed specifically for the purpose, and then by chance, to the so-called "personal" microcomputers. We were surprised to find that for the price of a conventional signal averager (1) one could obtain a stand-alone microcomputer system featuring large internal and external memory capacities, a CRT terminal, circuits necessary for creating instrument interfaces and a high level language. Clearly, a computer of this type may function not only as a signal averager but also as a general purpose data processor (2).

We then chose to purchase a microcomputer with the belief that it would have many applications beyond its intended signal averaging role, an expectation which was quickly confirmed. Alternative NMR data acquisition modes and tandem programs which extract linewidth information from signal averaged spectra were soon created (1,3,4). The system expanded progressively, and it now consists of the components and instruments shown in Figure 1.

The instruments serviced by the microcomputer fall

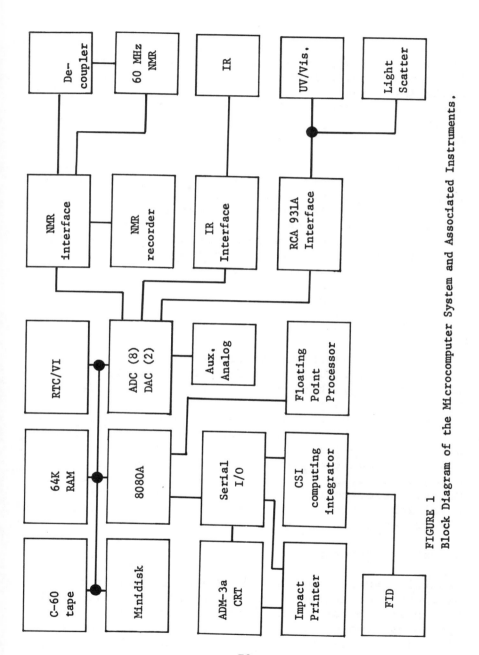

FIGURE 1
Block Diagram of the Microcomputer System and Associated Instruments.

into the moderate usage category, and each may be operated manually if the computer is not required.
Thus, by grouping instruments used once or twice
daily (or perhaps less) around a single microcomputer
one realizes more cost effective data processing. In
addition, some applications have been strictly off-
line, e.g., the data were entered at the CRT terminal.
   Personal computers are well suited for the chemical
laboratory since the interface problem has been much
simplified by general purpose digital and analog in-
put/output devices. This paper includes interconnecting circuit diagrams and driver listings appropriate for some of the instrument interfaces. An interface consists of both hardware and software. The
term "driver" refers to program features which move
information through the interface.

SYSTEM HARDWARE AND SOFTWARE

Microcomputer

   The Altair 8800b was selected for the applications
discussed here. It is based on the 8-bit Intel 8080A
microprocessor, which has access to 62K (K=1024) of
dynamic RAM memory. External magnetic memory consists of a C-60 tape cassette unit and two minifloppy
disk drives. A PROM loader located above 62K greatly
simplifies the initialization of DOS and minidisk
BASIC. In addition, the system may be operated with a
hardware floating point processor, the North Star
FPB-A. Unfortunately, the latter device compromises
total memory since it will not function with the upper
16K of memory in place.

Interface Components

Digital: RS-232 serial data may be linked to the
microcomputer through Altair's 88/2SIO standard digital interface card, which supports two ports. The
system has two of these, thus providing for three
digital channels in addition to the one always occupied by the CRT console.

A line printer is also linked to the CRT channel
(Figure 1) and can be turned on or off at will. The
88/2SIO ports may be configured for various data formats (parity, stop-bits, etc.) merely by sending appropriate statements to their control registers; no
wiring changes are involved.

Analog Conversion: Analog signals ranging between 0
and 10 volts may be converted to integers ranging,
respectively, between 0 and 255 (i.e., 8-bit precision) using the Altair 88-ADDA printed circuit
board. The reverse conversion is also possible, as
shown in Figure 2. Each 88-ADDA has two digital to

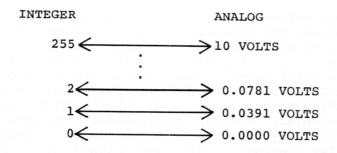

```
        INTEGER                      ANALOG

          255 <───────────────> 10 VOLTS
                        .
                        .
                        .
            2 <───────────────> 0.0781 VOLTS
            1 <───────────────> 0.0391 VOLTS
            0 <───────────────> 0.0000 VOLTS
```

FIGURE 2
Bidirectional Integer/Voltage Conversions in an Idealized Altair
88-ADDA Analog Interface. Voltage Calibration will vary from
circuit to circuit.

analog channels and eight multiplexed analog to digital channels. Obviously, a circuit of this type can
support several instruments. There are two such
88-ADDA units in the microcomputer described here.

Adaptation Circuits: Analog and digital devices of
the type cited have been available for several years
and are relatively inexpensive (less than $400). This
is true not only for the Altair microcomputer but also
for many of the others. Linkage of the 88/2SIO to a

serial data source is merely a matter of wiring (or
buying) a standard RS-232 interconnecting cable. On
the other hand, adaptation may be necessary in the
case of an analog device. For example, an instrument
which produces a signal ranging between 0 and 1 volt
will need some kind of intervening amplifier with a
gain of 10 if the full precision of the 88-ADDA is to
be used. An appropriate device for this purpose is
the 741 operational amplifier, a small scale integrated circuit priced at about $0.50. A single low
voltage power supply costing less than $400 will ope-
rate dozens of the low power 741 circuits.

Software

    System software for program development and opera-
tion include the following: 1.) Three versions of
interpreter BASIC (one on C-60 tape and two on minidiskettes); 2.) A locally-developed program which
operates octal coded machine language subroutines
from BASIC; 3.) A minidisk version of Microsoft's
DOS/FORTRAN-80 compiler; and 4.) A FORTRAN compatible assembler. One may thus trade off between program efficiency and ease of development. Minidisk
data files created in BASIC are FORTRAN compatible,
i.e., the DOS monitor will mount these disks and the
information is accessible to FORTRAN programs.

SPECIFIC DATA CHANNELS

RCA 931A Interface

    The microcomputer has been interfaced to a single-
-beam MPI 1000 UV-Visible spectrophotometer, specifically to its RCA 931A photomultiplier as shown in
Figure 3. The cascade photocurrent flows from the
anode to ground through load resistor R_6, producing a
negative voltage at the first 741 operational amplifier. Resistors R_4 and R_5 set a gain of 40, that is,
the factor $(1 + R_4/R_5)$ is 40 (5). The second amplifier inverts the signal to a positive polarity without

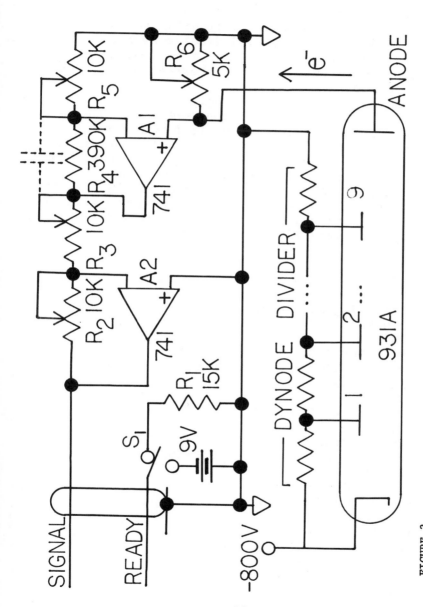

FIGURE 3
RCA 931A Photomultiplier Interface Circuit.

changing amplitude, i.e., $R_2/R_3 = 1$. Thus, the 0 to
-0.25 volt input of R_6, when multiplied by the circuit, becomes a 0 to 10 volt signal, and thus matches
the range of the analog to digital converter. To be
practical the signal plus noise must be slightly
smaller than 10 volt to avoid overrunning the range
of the 88-ADDA.

The handshake switch, S_1, determines a control voltage at a different analog input channel. One setting of the switch corresponds to 0 volts and the
other to 9 volts. Lines 110-150 of the data input
program (shown below) comprise a holding loop. The
command at Line 110 instructs the 88-ADDA to digitize
the voltage at the handshake channel and INP(57) at
Line 120 recovers the result. The computer will remain in the loop until S_1 selects 9 volts (Figure 3).
This is done every time a sample or blank is to be
read.

```
1ØØ   Y=Ø
11Ø   OUT59,1Ø
12Ø   IFINP(57)>1ØTHEN14Ø
13Ø   GOTO11Ø
14Ø   N=N+1
15Ø   IFN>1THEN21Ø
16Ø   S=Ø:FORJ=1TO1ØØ:OUT59,8:S=S+INP(57):NEXTJ:
      T1ØØ=S/1ØØ
18Ø   OUT59,1Ø
19Ø   IFINP(57)<1ØTHEN11Ø
2ØØ   GOTO18Ø
21Ø   REM---CALCULATION ROUTINE---
22Ø   S=Ø:FORJ=1TO1ØØ:OUT59,8:S=S+INP(57):NEXTJ:
      TS=S/1ØØ
24Ø   T=TS/T1ØØ:PCTT=T*1ØØ
25Ø   R=1/T:AB=LOG(R)/2.3Ø259
26Ø   IF AB>2.4THEN DD=1
27Ø   Y=Y+1
28Ø   REM---PRINT-OUT ROUTINE
29Ø   IF DD=1 THEN 32Ø
3ØØ   PRINT Y,TS,PCTT,AB
31Ø   GOTO 34Ø
```

```
32Ø  PRINT Y,"MEANINGLESS READING.  ABSORBANCE AP-
     PROACHES INFINITY"
33Ø  DD=Ø
34Ø  A(Y)=AB
35Ø  N=Ø
36Ø  GOTO18Ø
```

A blank is always read prior to each sample in order to establish a 100% T reference, and Line 14Ø is a counter used to distinguish between blanks and samples. When N=1, the sample data are processed at Line 21Ø. If N=Ø, the signal output of A2 is read as a blank. Line 16Ø acquires blank data and is the equivalent of a filter capacitor (Figure 3) since it averages 100 readings. The mean photometric intensity of the blank is thus variable T1ØØ. Similarly, the intensity of an absorbing sample is obtained as variable TS in Line 22Ø.

Fractional transmittance, from which absorbance is easily calculated, is simply TS/T1ØØ. One will note that the T1ØØ and TS values are not corrected for dark current since the latter reading is zero in a properly balanced circuit. Also, the number of readings averaged may be increased at will merely by changing the loop size in statements 16Ø and 22Ø. By sampling a slightly noisy signal and averaging it, a precision somewhat greater than the original 8 bits can be obtained. This may be envisioned by considering the result of adding two 256 level (8-bit) digitizations, which produces the effect of 512 levels. By adding a large number of readings, random errors due to photomultiplier noise and the digitization process itself may be partially compensated. If the photomultiplier output is filtered one must be sure that the interval between conversions is greater than the time constant of the circuit in order to receive a statistically valid sampling of data.

Lines 18Ø-2ØØ form the second holding loop, and the computer will remain in this loop until the switch is toggled back to the 0 volt position, preparing the system for a new reading.

Lines 21Ø-36Ø calculate the %T and absorbance and give a print-out of the results. Also, the statement at 34Ø saves the sequence of measured absorbances as subscripted variable A(Y), which may be filed on disk and passed to a data reduction program.

Infrared Interface

The infrared data channel is based on a circuit similar to the one discussed in the previous section. Resistors R_4 and R_5 shown in Figure 3 are set for a gain of 10 since our Beckman Acculab 10 produces a 0 to 1 volt signal. Neither R_6 nor the 941A are necessary in this arrangement, and the output of the first amplifier goes directly to the 88-ADDA, i.e., the input signal is positive. The operating program must be able to read a 1024 point array of spectrum amplitude during the scan interval, allowance being made for wavemeter pauses. Two channels are necessary, one for the signal voltage and a second for the synchronization pulses generated by the Acculab 10 wavemeter.

Correlation NMR (Narrow Scan)

Interface Hardware for correlation NMR is shown in Figure 4. Output channels A and B on the 88-ADDA Board step the voltage, creating a ramp. The offset voltage applied to summation point (X) determines the field position at start time. Operational amplifier D inverts the signal that is sent to the NMR, producing an upfield sweep. A 512 point resolution is obtained using this two channel combination, but greater resolution (e.g., 1024 points) requires synchronizing an integrator ramp (1) with an input loop. The net gain of C and D create a 2 ppm (120 Hz) field sweep. Amplifier S routes actively filtered spectrum amplitude information to the 88-ADDA channel Ø input.

Graphic playback is achieved through operational amplifiers A and B, shown in Figure 4. Amplifier A controls horizontal carriage movement. The NMR pen carriage is driven by a stepping motor which receives

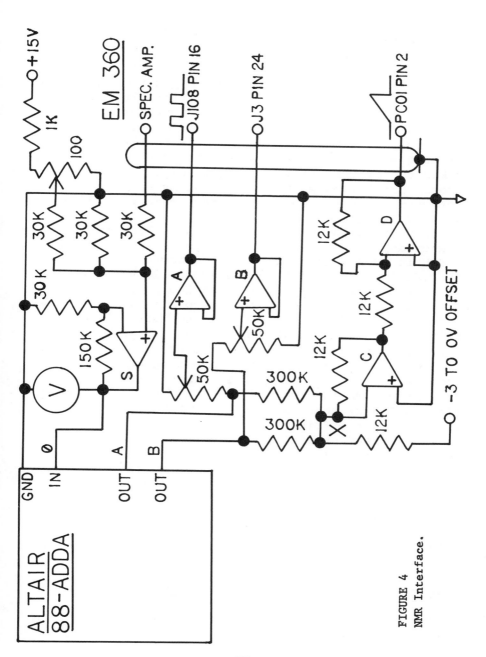

FIGURE 4
NMR Interface.

67

clock pulses from a frequency divider series. Two
pulses at the input (J 1Ø8 pin 16) advance the motor
one step. During computer operation the stepping motor
is driven by amplifier A using the program generated
rectangular waveform shown at the right in Figure 4.
Amplifier B controls vertical pen movement and thus
conveys spectrum amplitude information (1).

Correlation NMR is achieved by a FORTRAN source
listing much too large to describe here. Our version
makes use of the Cooley-Tukey (6) algorithm and the
cross correlation method developed by Dadok and
Sprecher (7). The spectrum array ordinarily contains
512 points, although 1024 is possible.

In fast sweep NMR the maximum possible ringing frequency depends on sweep width. For example, if sweep
width is 60 Hz, the maximum would be 60 Hz, and the
Nyquist frequency (f_n) is therefore 120 Hz. The sampling rate should not be less than 120 points s^{-1} under
such conditions. Therefore, sweep time should be less
than 4.3 s, and the digitizer input must be low pass
filtered to roll off at about 60 Hz. In biological
line-broadening studies, line widths are on the order
of $\Delta v_{1/2} = 1$ Hz or greater, and ring down time is
about $2/(\pi \Delta v_{1/2})$ or 0.64 s. This is well within
the sweep time interval of $1.3 < t < 4.3$ s. In practice, a 120 Hz spectral width may be scanned in about
2.5 s without difficulty.

The program creates a dead time or delay between
sweeps, which allows flyback ringing to decay while
the newly acquired spectrum is accumulated, or added,
to memory in the averaging process (integer array to
floating array). This is shown in Figure 5. In one
hour, about 1000 scans can be accumulated. Field lock
is essential. The number of scans that could be accumulated without over-running the mathematical precision is 39,000.

Pyrolysis Gas Chromatograph/Computing Integrator

The pyrolysis gas chromatography, PYGC, System consists of a pyrolyzer (Chemical Data Systems, Pyroprobe

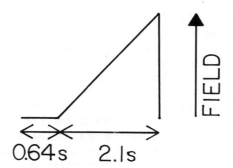

0.64 s   2.1 s

FIGURE 5
Output Waveform at Amplifier D (in Figure 4) shows a 0.64 s dead
time preceding a 2.1 s field sweep (then the cycle repeats). The
field sweep interval may be varied by changing a delay loop
counter entered during program operation.

100 with both platinum ribbon and platinum coil/quartz
tube heating elements) connected to the inlet of a gas
chromatograph (Varian Model 1520-1B) with dual flame-
-ionization detectors and a linear temperature programmer.   A SCOT column (Perkin Elmer) containing Car-
bowax 20M is used in this work.
    The output from the chromatograph electrometer is
sent to the 0 to 1 volt input of a Supergrator 2 (Columbia Scientific Industries, Austin, Texas), which is
a microprocessor based programmable computing integrator.   The standard Supergrator 2 computing integrator, CSI, contains 1024 words of Random Access
Memory, of which 648 are reserved as work area for
signal analysis, leaving 376 words for raw data and
programs.   Our particular CSI contains an additional
1024 words of RAM, providing added capacity for accom-
modating complex pyrolysis chromatograms.   The purpose
of signal analysis as performed by the CSI is to distinguish chromatographic peaks from baseline noise,
identify baseline segments, recognize small peaks and
any unusual abnormalities such noise spikes or sudden
baseline shifts, and to properly account for these
conditions in calculating peak areas or peak heights (8).

Baseline analysis is done automatically or can be forced in various ways (force baseline at a designated time, at nearest valley, project a horizontal baseline forward or backward, etc.) through programmable timed events. After corrections are made for baselines, tailing peaks, noise spikes, etc., the area and/or height and retention time of each peak is reported. The percentage of each component in the mixture can be determined using any of three calculation methods: 1.) Normalization; 2.) Internal standard, or; 3.) External standard. When the chromatographic run is terminated, the area/height, and retention time of each peak is stored in memory. The area of selected peaks can be deleted if desired, and percent concentrations can be recomputed. Additional programs to calculate results for method Two or Three (cited above) can be called into operation.

The computing integrator has been interfaced with the Altair 8800B microcomputer to permit additional data analysis. The CSI contains a printed circuit board Type SIOE (RS-232 type) serial interface set to operate at 300 baud, which is wired into an Altair input serial interface (88-2SIO), also set at 300 baud. A detailed schematic of the necessary wiring interconnections is shown in Figure 6. Since the CSI system has the ability to store area and retention time data in its memory, it can serve as a data buffer and can be activated at will to transmit data into the Altair microcomputer. To send information from the CSI one first loads and executes the microcomputer operating program, with the CSI printer turned off. The CSI "recompute" feature is then activated by sending an ASCII(#) from the computer to the CSI, or by pushing the recompute key on the CSI operating keyboard.

The algorithm listed below contains an input loop and a decoder for extracting the arriving numerical information from the ASCII array X(I). Upon completion, arrays RT(J) and AREA(J) contain retention times and relative areas respectively, while S(J) holds the CSI-determined peak numbers. This extract of the operating program will acquire a pyrolysis data set in

roughly one minute. Most of the delay is inherent in
the 300 baud rate of the connecting RS-232 link.

```
1Ø    REM-CSI/ALTAIR INTERFACE DRIVER
2Ø    DIM X(3ØØØ):I=1
3Ø    REM-CONFIGURE 2SIO PORT FOR CSI FORMAT
8Ø    OUT18,3:OUT18,9
1ØØ   IF(INP(18)AND1)=Ø THEN1ØØ
11Ø   X(I)=INP(19)
12Ø   I=I+1
13Ø   GOTO 1ØØ
```

Note that the 88-2SIO port must use octal addresses
Ø22 (control) and Ø23 (data); these appear as their
decimal equivalents 18 and 19 in the program. When
the CSI indicates that it has returned to command, the
microcomputer is given a Control-C. One then types
"GOTO 32Ø" at the terminal.

```
32Ø   DIM S(1ØØ),RT(1ØØ),AREA(1ØØ)
33Ø   J=1:Z=Ø
34Ø   I=4Ø3
35Ø   I=I+1:GOSUB53Ø:Y=Y*1Ø:GOSUB65Ø
36Ø   I=I+1:GOSUB53Ø:GOSUB65Ø
37Ø   S(J)=Z:Z=Ø
38Ø   IFS(J)=ØTHEN77Ø
39Ø   I=I+2:GOSUB53Ø:Y=Y*1ØØ:GOSUB65Ø
4ØØ   I=I+1:GOSUB53Ø:Y=Y*1Ø:GOSUB65Ø
41Ø   I=I+1:GOSUB53Ø:GOSUB65Ø
42Ø   I=I+2:GOSUB53Ø:Y=Y*.1:GOSUB65Ø
43Ø   I=I+1:GOSUB53Ø:Y=Y*.Ø1:GOSUB65Ø
44Ø   RT(J)=Z:Z=Ø
45Ø   I=I+4:GOSUB53Ø:Y=Y*1Ø:GOSUB65Ø
46Ø   I=I+1:GOSUB65Ø
47Ø   I=I+2:GOSUB53Ø:Y=Y*.1:GOSUB65Ø
48Ø   I=I+1:GOSUB53Ø:Y=Y*.Ø1:GOSUB65Ø
49Ø   I=I+1:GOSUB53Ø:Y=Y*1E-Ø3:GOSUB65Ø
5ØØ   AREA(J)=Z:Z=Ø:J=J+1
51Ø   I=I+9:II=I:GOTO 35Ø
53Ø   Y=VAL(CHR$(X(I))):RETURN
65Ø   Z=Z+Y
66Ø   RETURN    (Data reduction begins at 77Ø)
```

```
DB-255                    CABLE (25 ft)              10-pin MOLEX
PCB SIOE PORT                                        ALTAIR 88-2SIO
(CSI REAR PANEL)                                     SECOND PORT

                                                     (THE MOLEX IS ON
FORMAT: 7 DATA                                       THE INTERFACE
BITS; 1 STOP BIT;                                    PC BOARD)
EVEN PARITY


 2— SERIAL DATA IN ———————— BROWN ———————— TRANSMIT ——8

 3— SERIAL DATA OUT ——————— RED ——————————— RECEIVE ——7

 6— DATA SET READY ———————— ORANGE ———————————————————9

 5— CLR SEND ——————————————— YELLOW —————————— CTS ——1

 8— CARRIER DET ———————————— GREEN —————————— DCD ——2

 7— SIGNAL COM ————————————— BLUE ————————— GROUND ——4

20— TERM READY ———————————— VIOLET —————————— RTS ——3
```

FIGURE 6
Altair 88-2SIO/Supergrator-2 Pin to Pin Wiring.

OPERATING CHARACTERISTICS

On-Line Instruments

UV-Visible Spectrometer: The 931A photomultiplier
channel originated in a desire for on-line colorimetric enzyme assays that would reduce errors, require
less manipulation, and allow data to be saved in a
form which would permit the merging of results from a
variety of assays. The interface fits these requirements and attains a precision comparable to a Beckman
DB-G. Programs which operate this instrument print
tabulations in real time and file data sets on minidisk. Thus far we have used only the visible part of
the spectrum, but measurements in the ultraviolet
would only require adding an appropriate UV source.
The 931A housing may also be attached to a laser light
scattering goniometer (Figure 1).

Infrared: The infrared spectrophotometer is being used
to evaluate the practicality of a microcomputer-based
poison identification system. A program which follows
the data input automatically locates and quantifies
the major absorption peaks, creating a tabulation of
absorption wavelengths and normalized amplitudes (9).
The results are then compared with a file of knowns
using programs developed for chemotaxonomy. Reproducibility problems are similar to those encountered in
chemotaxonomic GLC. (See Pyrolysis GLC for Chemotaxonomy.)

Fast Scan 60 MHz ^{1}H-NMR: In biological applications,
correlation NMR is a sensitive and selective method for
dealing with the frequently large amplitude differences
between the solvent resonance (i.e., HDO) and a resonance of interest. By choosing a scan which misses the
solvent resonance this problem is circumvented. Pre-
-irradiation methods used in pulsed FT NMR can lead to
saturation transfer effects which confound the spectra
of macromolecules (10). The choice of a limited region,
as opposed to the whole spectrum, is not unrealistic

since in biological interaction studies one often focuses attention on selected groups within a biomolecule or drug.

Computation times and costs projected for correlation NMR using different configurations are shown in Table I. Configuration 1 times were obtained from the existing FORTRAN-80/8080 combination. The remaining entries are good estimates based on CPU clock rates. The table shows the economic qualities of the microcomputer (FORTRAN-80/8080), and it should be noted that the system is easily expanded to 1024 point correlation in the 64K memory environment. In the latter case execution time increases by a factor of about 2.3.

One of the problems of this modification is the low fidelity unfiltered EM-360 signal bandpass (Figure 7). This can be alleviated by the use of a state variable active filter, which enhances the higher frequency regions immediately preceding the stopband cutoff, permitting compensation without having to modify internal circuits. Figure 8 shows an uncorrelated scan of $CHCl_3$ (left) and its correlation product (right) obtained with the ALTAIR/EM-360 combination.

Pyrolysis GLC for Chemotaxonomy: Pyrolysis chromatograms are compared with a library of averaged profiles for a variety of organisms. The search strategy is similar to that used in a reported off-line method (11) since all area and retention pairs of the input and a given reference are compared for similarity within a defined statistical limit ($\pm$ 6.25%). If an area/retention pair agrees, a 1 is scored; otherwise, the comparison yields a zero. The index of similarity is thus taken as the sum of all pairwise agreements. Its value may range between zero and the number of profile peaks being compared. In using this method one will accept as the most probable identity of the organism the library example having the greater number of similarities.

The information contained in the measured index of similarity depends on the number of peaks compared. For this reason, it is desirable to involve _many_ peaks

TABLE I. Correlation NMR: Execution Times and Costs Based on 512 Point FFT's.

CONFIG.	FEATURES & COSTS	TIME FOR FFT & REORDER	TIME FOR CORRELATION	TIME FOR IFFT, REORDER & REALTRANS	TOTAL TIME
1	FORTRAN-80/8080 $6000	47.0 s	28.2 s	59.8 s	135.0 s
2	FORTRAN-80/Z-80# $6000 + $185	23.5 s	14.1 s	29.9 s	67.5 s
3	ASSEMB/FPB-A/8080* $6000 + $259	4.7 s	2.82 s	5.98 s	13.5 s
4	ASSEMB/FPB-A/Z-80* $6000 + $444	2.35 s	1.41 s	2.99 s	6.75 s

#The Z-80 CPU is twice as fast as the 8080 (4 MHz clock). Since it has
the 8080 instruction set, this is a simple hardware replacement.

*These are conservative estimates.

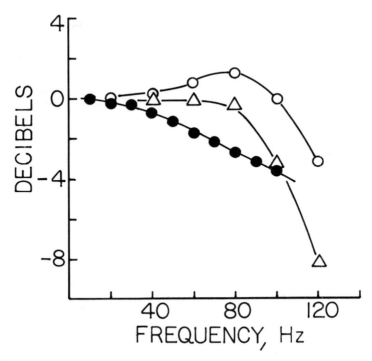

FIGURE 7
EM-360 NMR Bandpass Characteristics shown with Active Low-Pass
Filter Response Curves. Legend: ●, EM-360 response; △, Butter-
worth response; ○, state variable response for alpha = 1.

in the comparison. It is impossible to do this manually. In conducting the similarity comparison the retention times and areas received from the CSI are first
re-coded in a normalized two byte floating point representation (one byte exponent, one byte mantissa,
precision to 0.8%), using a small algorithm in BASIC,
and the results are POKED in a region of memory beginning at $25,500_{10}$ which is adjacent to the library references. As an example, a retention time of 26.5
minutes would code as mantissa byte = 212_{10}, exponent
byte = 197_{10}. Control is then transferred to an assembly language program which begins at memory location 25000_{10}.

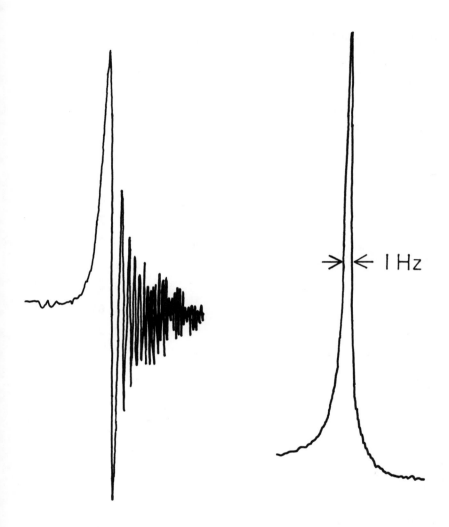

FIGURE 8
Correlated (right) and Uncorrelated (left) Scans of $CHCl_3$
Using the Altair/EM-360 Combination. The 1 Hz linewidth
is a result of mathematical convolution, which improves
the signal to noise ratio.

TABLE II.   Reproducibility of <u>Bacillus</u> <u>cereus</u>
Pyrolysis Chromatograms.

PEAK NUMBER	RETENTION TIME, MINUTES	PERCENTAGE PEAK AREA
1	1.00 ± .02	76.2 ± 3.9
2	1.70 ± .04	0.59 ± 0.53
3	1.96 ± .03	6.12 ± 0.90
4	2.34 ± .10	3.92 ± 0.82
5	2.80 ± .09	0.70 ± 0.16
6	3.31 ± .15	3.20 ± 3.1
7	5.86 ± .30	4.00 ± 1.4
8	12.75 ± .56	0.45 ± 0.28
9	13.34 ± .44	2.62 ± 0.51
10	14.74 ± .48	0.79 ± 0.13
11	15.31 ± .50	0.48 ± 0.05
12	16.78 ± .52	0.48 ± 0.28
13	17.81 ± .49	0.62 ± 0.28

The latter uses minimal arithmetic and examines for
all possible pairwise similarities between the new
chromatogram and each reference in the library.
Typical pyrolysis chromatograms of microbes may
show 30-60 peaks. Generally retention time reproducibility in replicate runs is achieved through careful
control of carrier gas flow rate and temperature programming conditions.   The data in Table II shows average retention times and percentage areas of thirteen
peaks from six pyrochromatograms of <u>Bacillus</u> <u>cereus</u>.
It is observed that the retention time and area reproducibility of the peaks varies widely.

The Altair microcomputer is an ideal companion to
the CSI as it can be used to calculate relative retention time, one-peak normalized retention time (12), or
two-peak normalized retention times for the various
peaks. The latter quantities are useful to correct
for variations between various runs. Similarly, peak
areas of replicate runs can be averaged or manipulated, after which classification programs of the type
discussed above can be called into use to compare unknown pyrochromatograms with references stored in a
classification library. The use of computers for such
comparisons seems to be the only hope for establishing
PYGC as a viable technique for the identification of
microbes. The lack of reproducibility of PYGC has been
recently pointed out by two workers (13,14). Blomquist
(14) and co-workers have suggested that the situation
with PYGC is not hopeless if the variation between the
chromatograms in different runs is somewhat systematic.
These investigators have used a pattern recognition
approach in an attempt to develop a classification for
Penicillium fungi (15). They describe their results as
an illustration of the possibility of using a model of
reproducibility based on principal components analysis.

Faster FORTRAN-80

The throughput of a minicomputer or a larger frame
machine may be increased considerably by adding a
sophisticated floating point array processor (16). In
a modest but similar way the number handling ability
of a small system might also be improved by including
a serial floating point unit. One example is North
Star's FPB-A, an arithmetic processor (costing $259 in
kit form) which executes single precision multiplica-
tions in an average time of 55 μs. This interval is
less than the time required to move two four-byte coding units into the processor and the four byte product
out (about 120 μs.). It is also much shorter than
the execution time of a floating point subroutine in
8080 BASIC or FORTRAN. The latter software modules
consume milliseconds, and in the case of FORTRAN-80,

floating point arithmetic becomes the bottleneck to throughput. We were thus tempted by the prospect of writing subroutines which will permit the operation of the FPB-A through FORTRAN-80.

Table III describes some of the FORTRAN-80/FPB-A subroutines developed to date. The full potential of this extension is realized only when "encode" and "decode" are used sparingly relative to the number of calculations. Also, one must be careful to avoid mixed mode conditions due to the peculiar binary coded decimal (BCD) format of FPB-A numbers, e.g., the latter cannot be combined with IF, +, -, /, *, or any library function. The 4 in REAL*4 and BCD*4 means that the number storage unit contains four memory bytes. Three are used to represent the mantissa and the fourth codes the exponent. REAL*4 and BCD*4 are not quite equivalent because the latter's mantissa is less densely packed (two decimal digits per byte) while its exponent spans a wider dynamic range (10^{+63} to 10^{-63}). The subroutines presented in this article could be modified to use the FPB-A's extended precision feature.

Algorithms which operated the FPB-A through FORTRAN subroutines were found to execute in as little as one--seventh the time of equivalent versions which used only the conventional (software) arithmetic. The limiting throughput of the 8080/FPB-A combination performing intense sequences of multiplication and division is about 5000 floating points per second, based on timed assembly programs. FORTRAN-80 itself will attain about 600 floating points per second.

Maintenance

In two years of operation, half a dozen equipment malfunctions have occurred. Of these, only one was an actual component failure (an operational amplifier on an 88-ADDA interface board). The rest were due to poor contact at connection points in the power supply, along the S-100 bus, and at the interface connections. All of these were hardly more than nuisances and were repaired by ourselves. Based on this experience, it

TABLE III. FPB-A Arithmetic Subroutines.

NAME	SUBROUTINE FORM	MEANING
Processor INITialization . . .	PINI.	Prepare for activity (use once at program's beginning)
Processor FINDer$^\phi$	PFIND(AD,x) . . .	Find the ADress of x (AD is INTEGER*2; x is BCD*4 or REAL*4
Processor ENcoder.	PEN(INT,BCD). . .	INTEGER*1 to BCD*4
Processor DEcoder	PDE(BCD,REAL) . .	BCD*4 to REAL*4
Processor Addition	PA(x,y,z)	$x = y + z$
Processor Subtraction. . .	PS(x,y,z)	$x = y - z$
Processor Multiplication . .	PM(x,y,z)	$x = y * z$
Processor Division	PD(x,y,z). . . .	$x = y / z$
Processor COMparison . . .	PCOM(I,x,y) . . .	Is x less than, equal to, or greater than y? Respective values of I will be -1, $\emptyset$ and +1. I is INTEGER*1; x and y must be BCD*4.

$^\phi$ PFIND is useful in numerous ways; for example, recovering a number with a decimal exponent less than -38 or greater than +38.

81

appears to be good practice to periodically clean the card connectors and jacks with isopropanol and a non--abrasive cloth. One should use top-quality minidiskettes since the cheap versions are unreliable.

System Sharing

Two unrelated research projects have access to the system on a continuing basis, and other temporary users appear from time-to-time. All of the instruments are within feet of the microcomputer and their interfaces are permanent features. Therefore, it is unnecessary to build up and tear down the circuits in the fashion of the small systems which are moved about on wheels. Since all of the data channels are kept "hot", the matter of operating a specific instrument is simply one of loading and running the appropriate program.

Off-Line Applications

When the microcomputer is not being used for the applications described above, it is available for other computational activities. Locally, programs have been developed in BASIC for the statistical treatment of experimental data. For example, we are able to do linear least squares analysis in double precision including the calculation of slope and intercept variance, and the correlation coefficients. A program has been adapted to perform non-linear least squares analysis in double precision with the system. Other programs available for student use include: 1.) GRAPHYV, a program that permits use of the printer as a graphic terminal; 2.) ALPHA, a program that allows one to calculate the fraction of each species in a weak acid equilibrium system as a function of pH;   3.) GAUS/POIS, a program that fits the Gaussian curve or the Poisson distribution function to chromatographic peaks; 4.) ENZLB, a least squares analysis for Lineweaver-Burke enzymological data; 5.) CHART, a program which uses the NMR recorder as a general purpose graphics terminal.

These applications are given to illustrate other possibilities for utilizing the microcomputer. Ready availability of such a system in the chemical laboratory (as contrasted with a large computer remotely located and restricted) is not only a great convenience, but also is a visible inducement that encourages students and technicians to develop a working knowledge of programming techniques.

Supported by NIH grant #5 S06 RR 08003-09 and NASA grant #NGS-00-802. A FORTRAN listing for correlation NMR and drivers for the FPB-A arithmetic unit are available on request.

REFERENCES CITED

1. J.R. Wright, Rev. Sci. Instrum., 49(9), 1288 (1978).
2. K.L. Ratzlaff, Am. Lab., 10(2), 17 (1978).
3. N.A. Jefferson, et al., Physiol. Chem. Phys., 11(3), 54 (1979).
4. J.R. Wright and J.L. Robinson, J. Chem. Ed., 56(10), 643 (1979).
5. A.J. Diefenderfer, "Principles of Electronic Instrumentation", Chapter 9, W.B. Saunders Co., Philadelphia, 1972.
6. J.W. Cooley and J.W. Tukey, Math. of Computation, 19, 297 (1965); see also W.D. Stanley and S.J. Peterson, Byte, 3(12), 14 (1978); R.H. Lord, ibid, 4(2), 108 (1979).
7. J. Dadok and R.F. Sprecher, J. Mag. Reson., 13, 243 (1974).
8. "Supergrator-2 Operating Manual" and "Supergrator-3 Programmable Computing Integrator Brochure", Columbia Scientific Industries, Austin Texas.
9. P.F. DuPuis et al., Anal. Chim. Acta, 112, 83 (1979).
10. J.D. Stoesz, et al., FEBS Letters, 91(2), 320 (1978).
11. F.M. Menger, et al., Anal. Chem., 44, 423 (1972).

12. R.J. Jolley and C.D. Scott, J. Chromatog., 47, 272 (1970).

13. H.L.C. Meuzelaar, et al., in "Rapid Methods and Automation in Microbiology", H.H. Johnson and S.W.B. Newson (Editors), Learned Information (Europe) Ltd., Oxford pp. 225-230 (1976).

14. G. Blomquist, et al., J. Chromatog., 173, 7 (1979).

15. idem, ibid, 173, 19 (1979).

16. H.F. Davis, Indust. Res., 19(11), 82 (1977).

T. F. NIEMANN, M. E. KOEHLER,
and T. PROVDER

Microcomputers used as Laboratory Instrument Controllers and Intelligent Interfaces to a Minicomputer Timesharing System

The primary use of the computer system described in this paper is the automation of laboratory instruments. This function includes real-time data collection, data analysis, control signal output for closed-loop control of the instrument, and post analysis data reduction for calculations and report ing the results of the experiment. Table 1 contains the list of instruments we initially selected for automation. This diverse assortment of instruments means that the computer system has to be extremely flexible as each instrument requires its own set of programs.

TABLE 1

INSTRUMENTS SELECTED FOR AUTOMATION

Instron Tensile Tester
Photogoniometer
Ferranti-Shirley Viscometer
Mettler/Paar Densitometer
DuPont Thermal Analysis System
Dynamic Oscillatory Rheometer
Gel Permeation Chromatograph

One of our objectives is to have the chemist responsible for the instrument do the major portion of the programming. In particular, we want the chemist to develop the complex data reduction program used for the final data analysis. To achieve this objective we need a computer system which can support a high level language such as FORTRAN or BASIC and all of the tools required for efficient program development such as a text editor, debugging aid and high speed printer. Other requirements we set for the system include a capability for graphics and disk or tape storage for the programs and acquired data. To meet these requirements we designed a network consisting of satellite microcomputers communicating with a host minicomputer. Each instrument has its own microcomputer and shares the facilities of the minicomputer. The microcomputer is the interface between the instrument and the minicomputer.

Mini-Micro System Organization and Operation. Table 2 contains a detailed list of the minicomputer hardware and peripherals. The disk and tape drives, printers, plotters, and terminals are all connected to the minicomputer. The microcomputers access these devices through the minicomputer. The microcomputers and terminals are distributed throughout the Research Center and are connected to the minicomputer with twisted-pair shielded cable. Data transmission is serial asynchronous using twenty milliampere current loops. The DZ-11 multiplexor is used to interface both the microcomputers and terminals to the minicomputer.

TABLE 2

MINICOMPUTER HARDWARE

Digital Equipment Corp. PDP 11/34 CPU
124K Words Semiconductor Memory
DZ-11 Asynchronous Multiplexor (32 channels)
DL-11 Telephone Interface
3 RK05 Disk Drives (12.5 megabytes)
1 RK07 Disk Drive (28 megabytes)
1 Digital Pathways TCU-100 Clock/Calendar
1 Computer Labs T-9100 Tape System
2 Gould 5005 Electrostatic Printer/Plotters
1 Texas Instruments Omni 810 Serial Printer
1 DEC VT-52 Video Terminal
9 Hazeltine 1500 Series Video Terminals
1 Pro-Log M-900 Prom Programmer

The minicomputer operates in a general time-sharing mode using the RSX-11M operating system, and the microcomputers operate in real time. The functions of the minicomputer are to control all the peripheral devices and provide the chemist with general timesharing services. Program development is done on the minicomputer, and the final instrument data reduction programs run on the minicomputer. Also available on the minicomputer are program packages for statistical analysis and data base management. The microcomputers perform all real-time applications required for data collection and instrument control.

There are four stages in an automated instrument analysis which are shown schematically in Figure 1. In the first or dialog stage the technician or instrument operator initiates the analysis. The operator uses one of the remote terminals to communicate with the minicomputer. During this dialog the minicomputer asks the operator a series of questions about sample identification and the parameters required for instrument operation and data analysis. The minicomputer checks the status of the microcomputer, making sure it is turned on and is not busy with another analysis. When the dialog with the operator is complete the minicomputer sends all of the input data to the microcomputer. The microcomputer acknowledges that it received the data and turns on a status light at the instrument, indicating to the operator that it is ready.

FIGURE 1

AUTOMATED INSTRUMENT OPERATION

1. Dialog

    Operator⟷ Mini⟶Micro

2. Data Acquisition

    Micro⟷Instrument

3. Data Transmission

    Micro ⟶Mini

4. Data Reduction

    Mini⟷ Operator

The second stage is data acquisition. The operator initiates the second stage when he starts the instrument. During this stage the microcomputer acquires data from one or more detectors in the

instrument. If the microcomputer is controlling
instrument operation some analysis of the incoming
data may be required and the necessary control signals
output. The microcomputer stores the data required
for the final analysis in its memory. During data
acquisition the microcomputer is performing all
operations in real time. The minicomputer is not
involved in this stage. The microcomputer notifies
the minicomputer when data acquisition is complete.
    During the third stage the microcomputer trans-
mits the data it collected and stored in its memory
to the minicomputer. The minicomputer stores this
data in data files on one of the disks.
    The fourth stage is the final data analysis and
takes place in the minicomputer. This data reduction
is generally done by one or more FORTRAN programs
written by the chemist. Reports and plots are generated at this time. The operations during this stage
may occur automatically or may be controlled by the
instrument operator.

    Microcomputer Hardware. To the instrument opera-
tor the microcomputer is a box with a power switch and
a reset button. Inside the box are power supplies and
a card rack holding 4-1/2 x 6-1/2 inch circuit cards.
One of the primary benefits of using individual microcomputers for each instrument is that the hardware and
software in each microcomputer can be tailored to the
specific application. However, there are also very
definite advantages in standardizing the hardware and
software. Using standard hardware decreases the time
required for design and documentation, simplifies
maintenance and makes it possible to use standard
software routines. In our microcomputers we implement various hardware functions using standard circuit
cards. Whenever possible we use commercially available cards since these cards are already debugged and
well documentated. The necessary flexibility is
achieved by building the specialized functions for a
specific application on separate cards. Frequently
needed specialized functions are lamp and relay
drivers, signal level translation, and switch debounc-
ing.
    The microcomputers are based on the Pro-Log 8821
processor card which uses the Intel 8080A microprocessor chip. In addition, this card contains 1k bytes
of RAM memory and sockets for up to 4k bytes of EPROM

memory. The amount of memory in each microcomputer
is determined by the application.

Analog to digital conversion of the analog signals is accomplished using an Analog Devices RTI 1220-
12 card which contains a 16 channel single ended or 8
channel differential multiplexed 12-bit A/D converter.
In addition, we have designed and built a card containing an active filter and instrumentation amplifier
to condition the analog signal before it is passed to
the A/D converter.

The following functions are currently implemented
on two wire wrap circuit cards designed in our laboratory. However, equivalent functions are now available
from Pro-Log or Mostek, and our next generation microcomputers will use these commercial cards. Serial
asynchronous I/O for communication with the minicomputer is implemented using a UART. Two 20 milliampere current loops provide full duplex operation.
The BAUD rate is switch selectable. Parallel I/O is
used for switch sensing, status lights, digital control and TTL compatible instrumentation and is implemented using a Motorola 6821 peripheral interface
adapter. The Intel 8214 priority interrupt control
chip is used to provide eight priority levels of
vectored interrupts. A real-time programmable clock
is implemented using the Intel 8253 programmable
counter/timer chip. The clock is controlled by a
2 MHZ crystal oscillator. The outputs from the clock
are tied to interrupt requests. The clock is primarily used to control the data acquisition rate.

  Microcomputer Software. The software for the
microcomputers is currently written in assembler.
Since most chemists have no experience is assembler
programming, this software is written by the Computer
Center staff. The benefits derived from using standard hardware modules have already been discussed.
Using standard software similarly reduces program
development time and documentation requirements. A
standard subroutine only needs to be written, debugged, and documented once. Consequently, we are making
a determined effort to standardize the software in the
microcomputers as much as possible. This software
consists of a monitor, standard system subroutines,
and the application program. Using the monitor the
programmer can download a program from the minicomputer, single step through the program, examine and

change the contents of memory locations, and initiate
I/O operations. The monitor and application programs
use the system subroutines. All communication with
the minicomputer is handled by these subroutines.
There are subroutines for data conversion such as
hexadecimal to binary and for various hardware device
operations.

The application program contains the customized
software for the specific instrument application.
Even here standardization is possible using system
macros. For example, there are macros available to
the programmer for building the interrupt vector
table, setting the real-time clock rate, and programming the programmable interface adapter.

In order to achieve this degree of standardization extensive use of mnemonics is required. All
addresses for memory mapped I/O registers, I/O ports,
scratch pad memory locations and interrupt vectors
are assigned standard symbolic names. These locations
are always referenced using these symbols. In addition, symbols are assigned to offsets within tables,
bit positions for flags, and constant values such as
the amount of RAM memory. As each microcomputer is
assembled a hardware definition file is created which
defines these symbols with the unique values for that
microcomputer. Thus, even though the hardware addresses may vary from one microcomputer to another, the
monitor, system subroutines and application programs
can be assembled to run in a specific microcomputer
without having to change the source code.

All software for the microcomputers is initially
developed on the minicomputer timesharing system using
the standard DEC text editor, assembler and task
builder. The main advantage in using these DEC utilities is that they are generally much more powerful
than the utilities available for most microcomputers.
The first step is to generate a disk file containing
the source code. The source code is submitted to the
assembler along with a macro library which defines all
of the mnemonics for the Intel 8080 instructions and
the file which defines the microcomputer hardware.
The output from the assembler is a file containing
relocatable binary code. This file is given to the
task builder along with a symbol definition file which
contains the absolute addresses of all the system
subroutines. The task builder resolves all these
global addresses, relocates the main program to the

specified absolute starting address, and produces a
file containing the binary machine code. Finally,
this file is submitted to a program we developed which
downloads the code into RAM memory in the microcomputer or sends it to a PROM programmer where the
program is burned into an EPROM chip. This chip is
then installed in the microcomputer. While the program is being debugged it is generally easier and
faster to download directly to the microcomputer. For
production operation the program is permanently
installed so that the instrument operator only needs
to turn on the microcomputer and run the dialog operation on the minicomputer to initiate an analysis.

W. C. FISCHER, W. L. SECREST,
J. E. CASSIDY, D. P. RYSKIEWICH,
and G. J. MARCO   CHAPTER 8

Development of a Distributed Laboratory Data Reduction Network Using Desk-Top Computers

Introduction. The quantity of data from the instruments in
a modern laboratory increases with each technological advancement or governmental regulation. Laboratory oriented data
processing is a cost and time effective way to organize these
data for meaningful interpretation. Although the examples used
in this paper are specifically that of Metabolism data for
Agricultural Products, the concepts, strategies and hardware
would apply to any chemical, physical or biological laboratory.

The Environmental Protection Agency (EPA) has established
rigorous standards which must be satisfied in order to register
a chemical for use as a pesticide. The time required to
generate the data necessary to comply with these regulations is
presently estimated to be nine years from the original synthesis
to the appearance of the chemical in the agricultural market-
place. The biochemical, environmental and toxicological effects
of the chemical and its metabolites must be well documented.
For example, data for a single herbicide, Dual®, for use on a
single crop, soybeans, occupied a stack of paper 10' tall.
Data handling of this magnitude required the extensive use
of computers for collection, reduction and summarization of
results.

However, the use of a single, centralized general purpose computer to fill these computing needs, from data collection to summarization for decision making, is often impractical for the laboratory. Neither will stand alone small computers provide the capacity for large data bases. A data processing hierarchy is a logical and cost effective way to use the strengths of various size processors and avoid their weaknesses.

The Metabolism Group in the Biochemistry Department of CIBA-GEIGY's Agricultural Division has decided to use a network of Hewlett-Packard 9800 series desk top computers for the first echelon of a data processing hierarchy, i.e., data collection. A HP-9825A services 4 liquid scintillation counters, a Hewlett-Packard 1084B HPLC and two remote terminals. One terminal, an HP-2647A, has full graphics capabilities and 10K of BASIC programmable memory so that data can be preprocessed. The other terminal is a HP-9835A desk top computer. Data can be exchanged among these processors allowing various laboratory groups access to the entire data base. Sophisticated I/O in the desk top computers provides simultaneous service to all I/O devices. This desk top computer network provides a versatile, laboratory oriented data collection/reduction system which is compatible with interim size and large central corporate computers.

Objective. The Metabolism Section has the responsibility for supporting product registration with timely and accurate scientific data. Three general questions about pesticides in agricultural products and the environment must be answered: where did the chemical and its metabolites go, how much are there and what are the metabolites. Metabolism studies are generally done with organic chemicals labelled with radioactive carbon (^{14}C) which provides a rapid and convenient method for detection and quantitation of a pesticide and its metabolites. The liquid scintillation counter (LSC) is a preferred way to measure radioactivity. From 1/2 to 22 minutes of LSC time is generally required per sample vial with most samples being counted three minutes. Data from 300 to 600 vials are processed on an average day (Table I).

Table I.   Quantity of Liquid Scintillation Counter Data
Generated

Time: 1/2 to 22 minutes per sample vial.

Typical Daily Volume: 300-600 sample vials.

Maximum Capacity: 1,000 sample vials.

The LSC outputs per.se. are not useful (Figure 1). The
data must be corrected for background radiation and counting
efficiency. The result is an absolute number, disintegrations
per minute (DPM), which can be used for quantitation. Modern
LSC's which calculate DPM offer little advantage in this
application because of the many calculations needed for meaningful results. Aliquot size, moisture factor and specific
activity must be used to calculate parts per million (ppm)
of a chemical, partitioning characteristics, or print histograms
which graphically present chromatographic behavior (Figure 2).
These calculations are beyond the capability of turn key LSC
data reduction options.

```
195  1  002.00  A=000813.5(5.0%)  B=000913.5(5.0%)  (=000000.0(>20%)  S=0.616
195  1  002.00  A=0007f2.5(7.0%)  B=000879.5(5.0%)  (=000000.0(>20%)  S=0.617

196  1  002.00  A=003910.0(3.0%)  B=004367.0(3.0%)  (=000000.0(>20%)  S=0.620
196  1  002.00  A=004004.0(3.0%)  B=004418.0(3.0%)  (=000000.0(>20%)  S=0.619

197  1  002.00  A=000348.0( 10%)  B=000440.0(7.0%)  (=000000.0(>20%)  S=0.464
197  1  002.00  A=000326.5( 10%)  B=000464.5(7.0%)  (=000000.0(>20%)  S=0.464
```

FIGURE 1: Liquid Scintillation Counter Output

```
File Name: P335
Line Number                      Description            Entry
-----------                     -------------           -----
     7                          Background           -30.0000
     8                          First Sample          81.0000
     9                                                 2.0000
    10                                                 8.0000
    11                    Combustion Efficiency         0.9700
    12                      Specific Activity          25.0000
    13                     1st Aliquot Weight           0.2113
    14                     2nd Aliquot Weight           0.2105
```

```
File No. - P335
Program - corp
Employee I.D.: 13542  Notebook Ref: I
```

```
5/ 13/1980    Accountability/Date:
```

Specific Activity = 25.00

BKG	CPM1	CPM2	EFF	DPM	ALIQ	HARV	DPM/GRAM	PPM	PPM ERR
30	1730	1775	46.29	3721	0.2113	0.97	18154	0.327	0.011
30	1711	1674	45.78	3631	0.2105	0.97	17783	0.320	0.011

FIGURE 2: Technician Entries and Reduced Data

Data Reduction Options. Hand calculation (Table II)
was the original method of LSC data reduction. Data from 100
samples would typically require four hours for a minimum
amount of reduction. Early desk top programmable calculators
such as the Olivetti Programa 101 reduced this time to
approximately three hours for 100 samples.
   Both of the early methods of data reduction were manpower
intensive and, therefore, highly inefficient. The first cost-
effective method of LSC data reduction was time-share computer
service. Paper tape from the LSC's is read into the computer
from a terminal, the data are reduced and a formatted output
is obtained quickly and easily. The time needed to reduce the
data from 100 samples was approximately 30 minutes. The data
were routinely analyzed by the counting statistics of Currie
(1) so that the interpretations were more accurate. This is a
major advancement over previous methods of data reduction. The
time-share companies offer excellent technical support for
customers who lack in-house programming expertise. Time-share
service was generally reliable and moved Metabolism projects
rapidly.

   Table II. Time Required to Calculate 100 Samples

                 Hand Calculations: >4 Hours
                 Simple Programmable Calculator: 3 Hours
                 Time Share Computer Service: 30 Minutes
                 Directly Interfaced Mini-Computer: <2 Minutes

---

Despite the cost effectiveness of time-share service, large amounts of time were required to sort paper tape, enter it into a slow terminal and edit the entries. Downtime averaged about one day per month. There are many costs associated with time-share service which depend on use and data storage. A special low noise telephone line is also required. Variable costs are difficult to budget, therefore, policing the use of the service required approximately one man day per month. These weaknesses made the search for more cost effective method for LSC data reduction an ongoing task.

A dedicated mini-computer finally proved to be the most cost effective approach. After evaluating all the systems available in 1976, we selected the Hewlett-Packard HP-9825 and its versatile line of I/O peripherals. It was capable of handling the on-line data logging and data reduction but not so overpowered that dedicating it to a specific task would not fully utilize its power. It was also relatively economical.

The desk top computer has been a reliable, cost effective addition to the instrumentation in the Metabolism Section. By removing paper tape as a data transfer medium, approximately two man years of technician time has been re-directed to bench oriented work. Software changes can be made quickly, easily and cheaply. Most important, however, is that this basic data processing system can be expanded or changed to fit any particular need.

Hardware. The LSC's used by the Metabolism Section include two Beckman instruments, an LS-233 and an LS-255, and two Tracor Mark III's. The Tracor instruments have two standard communications outputs, RS-232-C and current loop, which will directly interface to the desk top computer. The Beckman instruments use a unique teletype driver which initially presented a problem. One of the criteria for the on-line data collection system was that it would be compatible with all departmental LSC's. This problem was solved with a William-Palmer Industries' interface which converts the Beckman output levels to RS-232-C. An especially attractive feature of this interface is that its microprocessor can be custom programmed by the manufacturer to further simplify interfacing. Therefore, on-line data collection became a relatively simple matter of software development.

The data collection hardware is based on a HP-9825A desk top computer (Figure 3). One megabyte of flexible disk memory, a printer/plotter and the necessary interface equipment complete the minimum LSC data reduction system. The LSC's and data system run virtually unattended, 24 hours a day, 365 days a year. Data are logged by the HP-9825 at an average interval of 45 seconds. Reliability was a major consideration when the equipment selection was made. Since installation in May, 1977, there have been less than two work days of cumulative downtime.

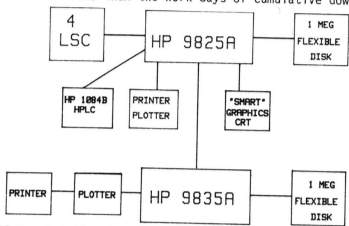

FIGURE 3: Existing Desktop Computer Network

The need for access to the HP-9825 and its data base increased rapidly as its usefulness became apparent. New hardware has been added to meet specific needs. These additions include a HP-2647 graphics terminal and a HP-9835A desk top computer which is supported by one megabyte of flexible disk memory, a 180 CPS printer and a four color X-Y plotter. All three processors are interfaced to each other so that data files may be exchanged electronically. At the present time, data processing is available at three different locations. All three processors give access to the same data base. This nucleus of a network of mini-computers provides access to data for a maximum number of people while retaining the reliability and flexibility of distributed processors.

Software. The buffered I/O feature of the Hewlett-Packard
desk top computers is what makes this network possible. The
input data go into buffers as they are received. When a
specific character or character count has been received, the
computer skips to a service routine. Simultaneous interrupts
are handled in the order of priority. No data are lost and
I/O operations appear to be handled concurrently.

The LSC service routines put the data into a standard
format and store it on a flexible disk. The last two passes
of data are stored, earlier data are discarded. The time
required to service an LSC is typically less than two seconds.

Technicians enter the information required for the
specific calculations, such as aliquot size, through the keyboard. The desk top computer treats the keyboard as another
interrupt device, so data can be entered without stopping the
LSC's. Responses are made to prompts on the computer's display.
New technicians are able to learn this interaction quickly.

Equipment which runs continuously must have provisions for
problems which occur when no one is available to correct them.
The HP-9825 has a provision for restarting after power failures.
Errors in LSC data are also handled by a service routine. The
error, data and line number in which the error occurred are
printed on the internal printer. This feature allows software
errors or hardware malfunctions to be rapidly traced and
corrected.

During the data reduction, the computer is continually
checking the data against tolerance and making judgements on
the results. These tolerances include the counting statistics
and a + 10% for the two data passes from the LSC. When this
+ 10% tolerance is exceeded, the computer assumes that the
latest pass is the most accurate and uses it twice. The output
is so noted and the technician knows to run the sample again.
These features are especially helpful for new or less
scientifically trained personnel.

Output. A hard copy is printed for each set of data
the technicians have entered (Figure 2). This output includes
"raw" data from the LSC's and technicians. Intermediate and
final results are printed making the calculations easy to check.
The real-time computer clock dates the output and provides a
space for the technician's signature and date. According to
EPA's proposed standards for Good Laboratory Practices, the

output would appear to qualify as "raw" data. Not only do the formatted outputs speed data interpretation and report writing, they will aid laboratory audits.

In addition to hard copy outputs, the desk top computer stores the information on a flexible disk in the same format as printed. Later, the data on the disk are sorted by sample and project. A single page is printed which contains all of the "raw" data, the calculations and a summary for each sample (Figure 4). By assigning data handling to computers, technicians are free to do bench-oriented work and scientists are free to plan and interpret experiments.

Sample # D313

Responsibility/Date

PPM	=	.654
Organic	=	6.6
Aqueous	=	47.8
Non-Ext	=	36.2
Total	=	90.6

Employee I.D.: 13542
Notebook Ref: I
6/ 14/1979

BKG	CPM 1	CPM 2	EFF	DPM	ALIQ	HARV	DPM/GRAM	PPM	PPM ERR
21	5669	5742	75.70	7509	0.1790	0.97	43279	0.650	0.017
21	5573	5724	75.81	7422	0.1748	0.97	43809	0.658	0.017

Employee I.D.: 13542
Notebook Ref: I
6/ 14/1979

Specific Activity = 30.00 ppm = 0.654 moist = 1.00
Total Weight =   5.0012

BKG	CPM 1	CPM 2	EFF	DPM	ALIQ	TOT VOL	TOT DPM	PERCENT	% ERROR
21	147	122	79.26	142	0.1000	10.00	14225	6.53	1.31
21	252	244	77.37	293	0.2000	10.00	14637	6.72	0.92
21	250	260	80.41	290	0.1000	25.00	72601	33.33	4.47
21	467	465	80.10	555	0.2000	25.00	69404	31.86	3.03
21	21	19	80.04	0	0.1000	25.00	(3830	(1.76	
21	30	23	79.37	7	0.2000	25.00	(1915	(0.88	
21	73	78	80.47	67	0.1000	50.00	33555	15.40	4.85
21	129	124	80.28	130	0.2000	50.00	32618	14.97	3.15

Employee I.D.: 13542
Notebook Ref: I
6/ 14/1979

Specific Activity = 30.00 ppm = 0.654 moist = 1.00
Total Wt =   5.0012   Res Wt =   2.2778

BKG	CPM 1	CPM 2	EFF	DPM	ALIQ	HARV	TOT DPM	PERCENT	% ERROR
24	3979	4033	75.44	5278	0.1604	0.95	78905	36.22	1.13
24	4056	4070	75.30	5363	0.1630	0.95	78900	36.22	1.12

FIGURE 4:   Summary of Sample Analysis

Expansion. Standardized electrical outputs are frequently available on modern laboratory equipment. Gas and liquid chromatographs are especially amenable to off-line data summarization. For example, a Hewlett-Packard 1084B HPLC which is frequently used for specific activity determinations has been interfaced to the HP-9825 which services the LSC's. Data from the counters are combined with data from the HPLC micro-processor and the specific activity of a compound is calculated. The data and calculations are returned to the printer on the HPLC. The chromatograms, calculations and results are, there-fore, included on a single sheet of paper providing a complete report (Figure 5).

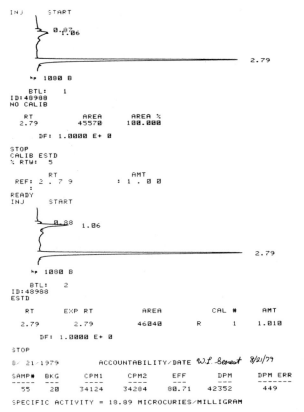

FIGURE 5: Specific Activity Determination

As more laboratory tasks are computerized, this network of desk top computers can expand to meet those needs. The cost is nominal and the risk of premature obsolescence is minimized. As needs change, the structure of the network can change to meet those needs. The network concept allows each of the distributed processors to do each of its assigned tasks. Data can be exchanged between processors to eliminate the errors inherent in human transcription. Therefore, obsolescence can be overcome with software, not costly hardware. The only basic requirement is a standard communications interface and modest software additions.

Larger central data processing systems, such as the corporate IBM-3033, are not well suited for handling the sporatic data from laboratory instruments. In the event of a malfunction, the entire data gathering/processing capability can be lost with a central unit. In a mini-computer network, a malfunction may temporarily inactivate a function but not the entire system.

However, desk top computers are not the answer to all data processing needs. In the laboratory they are ideal for data logging, reduction, and reporting. They are not well suited for operations on large data bases. Data collected by desk top computers in several departments can be sent to a larger computer for further processing and summarization. A strategy for accomplishing this within the Metabolism Section is with a data processing hierarchy (Figure 6). The desk top computer network, the computer in a HP-5985 GC-MS and a computer in a FT-IR are to be interfaced to a HP-1000 E computer. In addition to doing mathematical calculations and data analysis, the HP-1000 will store all of the data for a given project on a single disk. This will reduce the volume of the archives significantly. Other advantages of magnetically stored data include ease of copying data and computer assisted data retrieval.

This HP-1000 E is presently interfaced, by modem, to a central corporate computer, an IBM 3033 in Ardsley, New York. The corporate computer is capable of operating on very large data bases and providing long-term archival storage. It already has the software to aid with report generation. The Federal guidelines for data retention and retrieval make a large, central processor a practical medium for cost effective compliance.

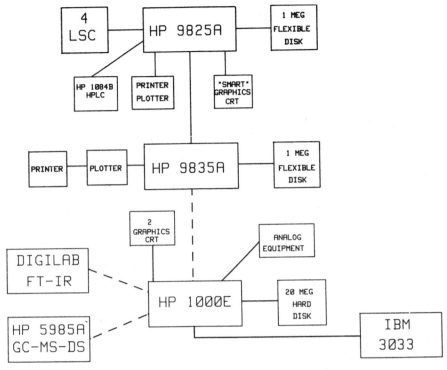

FIGURE 6: Expansion of Desktop Computer Network

Conclusions. Manpower intensive data and paper handling in an R&D organization is costly. Scientists and technicians are hired to solve scientific problems, not to be clerks. Cost effective data processing equipment is available to make these people more efficient and reduce errors. However, no single piece of data processing hardware best fits all the tasks which must be accomplished. A data processing hierarchy is a logical and practical way to overcome the weaknesses of any particular level.

Desk top computers are well suited to laboratory data collection. As laboratory instrumentation becomes more sophisticated with its ability to pre-process data, the desk top computers are especially useful for processing these data into summaries allowing scientific decision-making immediately available during the experiments.

A desktop computer network adds another dimension to data gathering by providing the flexibility of organizing data from distributed processors. Data from several locations can be shared and some equipment duplication can be eliminated. The computer becomes a tool, like a balance or chromatograph, and its use is just as easily learned. It is a part of the scientist's problem solving strategy controlled by the scientist, not by the availability of computer specialists and large multifunctional apparatus such as a central computer.

As data processing becomes more available at the bench level, we have observed that its use and usefulness increase. Its speed, availability and cost effectiveness have immediate benefits in developing new or improving old products by completing projects in as efficient a manner as possible. When the need for greater computing capacity develops, scientists are prepared to interact effectively with computer professionals. Data collected at the bench can be quickly summarized and transmitted to a large data base which can be accessed for both scientific and management decision making.

LITERATURE CITED

1.   Curie, L. A., _Anal. Chem._, 1968, 40, 586.

P. C. KAHN

A General Purpose Minicomputer
in the Biochemistry Laboratory

The use of computers by chemists is now sufficiently common that it needs little documentation.
One need only consult prior volumes of this series
(1, 2) to gain an idea of the broad range of research
areas within which computers are essential everyday
tools. The advent of the microcomputer (3), and
especially the development of the mass home market,
has so lowered the cost that microcomputer ownership
by individual laboratories is now possible and forms
the basis of this symposium. Cost is not the only
factor in the spread of these machines into the
research laboratory, however. Ease of use is at least
as important. The fact that chemists with little
prior computing experience can set up and run laboratory microcomputer systems is as much responsible for

*Paper of the Journal Series, New Jersey State Agri-
cultural Experiment Station.

today's discussion as is cost.

The same factors that have brought the micro-computer within the purview of individual laboratories have been operating throughout the computer industry. In particular, minicomputers have also become both cheaper and easier to use. This paper and the two that follow (4, 5) are a consequence of that trend. The work that we will describe was done by a faculty member familiar with higher level language programming on large machines (Fortran, PL/1) and by an under-graduate chemistry major who developed an interest in computing. Neither had any prior experience with assembly language, with systems programming, or with computer hardware, and neither is trained in elec-tronics. For neither of us was the development of the system our principal work.

The choice of a full fledged minicomputer rather than one or more micro systems or the use of time sharing on a big machine was dictated by the needs and background of the department. Our 14 faculty are distributed among biochemistry, microbiology and plant physiology. With the exception of the author none had any prior computing experience, and only an occasional graduate student found reason to become competent in programming. All the faculty felt that they could profit - some greatly - by having computing capability readily available. Unfortunately, and for reasons that are well known, the University IBM system, although having batch and time sharing terminals throughout the campus, is not generally seen by persons lacking computer experience as being easily available. In addition, online collection and permanent storage of data and automated control of apparatus are in-creasingly in demand, and remote time sharing systems cannot be used for such purposes. Online applications and ease of use required a system local to the department, and the varied needs of disparate research faculty justified substantial computing power.

In parallel with the purchase of the computer the department installed a Cary 17D spectrophotometer. The successor to the well known model 14, the new instrument produces BCD output of the optical density through a digital volt meter. Because many labora-tories would have use for computer collected absorption data, and because the availability of digital output eliminated the need for analog-to-digital conversion, we chose the interfacing of the Cary as our first project. This choice also provided the best justification for the use of state funds in the purchases.

The collection and storage on magnetic discs of Cary data are described in the following paper (4). That discussion includes design ideas aimed at making the data gathering procedures as general as possible for future application to other instruments. This paper gives a general overview of the system and its projected development, and in the third paper (5) interactive programs for the post-experiment analysis of spectroscopic data are described. Running throughout the series is a concern for general applicability. Wherever possible both software and hardware fulfill dual roles: an immediate research application and use as a building block of a larger projected application. In addition, we do as much programming in easily transportable Fortran as is practical in order that the work may be of use to other laboratories.

System Hardware. The system is built around a Computer Automation α-LSI Model 2-10/G computer. The word length is 16 bits, and the machine is equipped with 32K words of memory. With memory bank controllers it is expandable to 256K. Some of its characteristics are shown in Table 1. Although sold primarily as industrial controllers, these machines have a surprisingly powerful instruction set that makes them well suited for research use. Ours includes hardware instructions for integer multiply and divide, and versions with floating point hardware that are also somewhat faster than ours are now available. Floating point software is available for integer machines and includes double precision operations. The assembly language permits pseudo-operations that invoke the floating point software.

Reliability is good. In three years of heavy use we have had one hardware failure. That is notable, for air conditioning failures have several times sent the laboratory temperature to over 90°F, an overnight water leak from the floor above flooded the room, and there have been building-wide power failures. Our one breakdown was repaired by the manufacturer at a nearby site and the machine re-installed by us in one day at

TABLE 1

SELECTED MINICOMPUTER CHARACTERISTICS

MEMORY CYCLE TIME: 1200 NS
CPU CYCLE TIME: 300 NS
WORD SIZE: 16 bits
MEMORY SIZE: 32K words, expandable to 256K

REAL TIME CLOCK: Crystal derived,
                    1, 10, 100 msec/pulse, jumper
                    selected.
                    Pulse = interrupt to pre-set
                    location

REGISTERS: Four 16 bit, one 8 bit, three one bit
  16 Bit: A, X are full working registers. X usable
             as addressing index.
             I holds current instruction
             P holds address of next instruction
   8 Bit: System status
   1 Bit: EIN, interrupts enabled when set. Soft-
                  ware controlled.
             OV, Set upon arithmetic overflow, used in
                  shifts.
                  May be set/reset by software.
             BM, Byte mode. Software controlled.

MEMORY ADDRESSING
    Word or byte
    Scratchpad (1st 256 words addressable from any
                location)
    Relative
        256 words foreward, 255 words backward
        512 bytes foreward only in byte mode
    Indexed (X register added to instruction operand)
    Indirect (chained)
    Stack

INPUT/OUTPUT
    VIA registers under direct program control
    VIA DMA channel, several options including
          interrupt control.

107

a cost of $270.

A schematic of the full system is shown in Figure 1. Money problems prevented our obtaining the full system at the start, but the initial purchase, shown within the dashed line, was made with the organization of the full system in mind. Planning for future expansion should always be included in an initial purchase. The initial system in the figure was sufficient to allow Cary interfacing and regular use of data collection programs. The wiring of the single terminal I/O port was brought out to a panel on the front of the instrument rack. Switches permit selection of baud rate, interface type (RS-232 or 20 ma current loop), and terminal device. Transmission rates as fast as 9600 baud are standard on the computer, but we wired only those for which we would have use. All this is transparent to the machine, however, for it simply writes to or reads from the port address.

Note that at this stage the only printed output device was the teletype. Program development has been spurred by the subsequent addition of the line printer, which delivers approximately 80 lines per minute and which allows up to 132 characters per line on 8 1/2 inch wide paper.

The printer and a digital XY plotter are connected to the computer through the "distributed I/O system", which was purchased at the same time as the printer. The I/O system is a half card board in the computer chassis that carries four channels to each of which a small microprocessor specific to the device type of that channel is connected. Transmission rates for the RS-232 channels are jumpered by single wires on the half card board. The I/O system was installed by us in a day, although learning to program it took longer.

The 4800 baud RS-232 channel is intended for use with the Tektronix graphics terminal. This will allow the standard CRT terminal to be shifted from the teletype line to the position now occupied by the Tektronix. These changes await the alteration of our assembly language I/O drivers for Tektronix graphics. The alterations are minor but have been delayed by higher priority work.

The general purpose channel is logically identical to the digital I/O board with which the computer

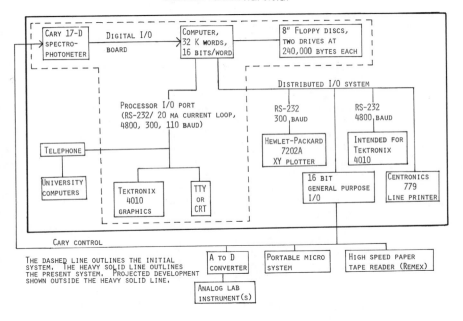

FIGURE 1

LABORATORY MINICOMPUTER SYSTEM

The dashed line outlines the initial system. The heavy solid line outlines the present system. Projected development shown outside the heavy solid line.

109

was originally purchased and to which the Cary is connected during data recording. The logic of our Cary drivers could thus be used with minimal modification. It is intended that this channel be switch selectable among four uses:

1.   Within the laboratory are other instruments that will be interfaced *via* an A to D converter. Chief among these is a spectrofluorometer.

2.   Apart from the teletype we have no paper tape input. Many laboratories in the department however, have instruments that produce extensive paper tape output which they would like to be able to analyze. Amino acid analyzers, scintillation counters and a microcalorimeter are typical. A Remex reader is on hand, and its installation is next in the hardware queue.

3.   The wavelength drive of the Cary is equipped with a stepping motor. By connecting the general purpose channel to it, we will be able to control the spectrophotometer from the computer. With the closure of the control and data gathering loop, repetitive scanning and a variety of complex experiments become possible.

4.   The department is now developing a portable microcomputer system which will include its own small floppy discs. Mounted on a cart, it will be moved about the building from experiment to experiment for data collection and control. Data analyses that exceed its capacity and graphical display of results will be performed by the minicomputer.

Of particular interest to our laboratory is the telephone connection to the University's large I.B.M. system. This is designed for the transfer of programs and/or data in both directions. Calculations that the minicomputer cannot handle can thus be carried out on the "number cruncher" with the results returned for graphical display, etc. The logic of the link is that normal communication is established between the I.B.M. system and the CRT terminal. When data are transferred in either direction, they are sent as 8 bit bytes. These are available for inspection by the

minicomputer. Specific sequences of characters
initiate any of the following operations regardless
of whether the sequence is sent by the University
system or typed at the terminal by the user:
1.   Transfer of a file from minicomputer floppy
      disc to the I.B.M. system or the reverse.
2.   Unloading or loading of all or part of the
      minicomputer's memory to or from the
      larger machine.
3.   Printing on the minicomputer's line printer
      of modest amounts of I.B.M. output.
Disposition at the I.B.M. end of the link depends
on the user's responses and may involve saving the
file directly as an I.B.M. data set for use by a batch
job or it may be saved within the text and program
editing system. Applications of the link
are described in reference (5).

    System Software. Although the present system
may seem elaborate, installing and wiring the hardware
was relatively straightforward. As with any computer
system the bulk of the development effort, instead,
is in the preparation of software. Our strategy to
minimize development time includes the following:
a.   As much programming as possible is done in
      Fortran rather than in assembly language.
      Fortran, of course, is easier both to write
      and to debug, and many functions, I/O
      especially, are handled by the operating
      system. We restrict assembly language to
      device dependent drivers or to the few
      functions that cannot be carried out in
      Fortran such as operation of the real time
      clock. Assembly language routines are
      Fortran callable.
b.   Subroutines in either language are designed
      to be as general as possible. Specific
      applications are built around a core of
      general routines with modular and easily
      replaceable routines for the current
      application. Several examples are described
      in the next two papers.
c.   Efficiency of both program execution and
      memory usage are often sacrificed deli-
      berately in favor of simplicity, the purpose

being to produce running code as rapidly as possible. We have two reasons for this. The first is that since the department had no prior history of computer development, it was expedient to show tangible results promptly. More importantly, by bringing up all parts of complex programs in simple versions, we find out where *all* the un-anticipated problems lie. As the resolution of one problem often affects that of another, we are able systematically to make improvements where they will do the most good. The result is that many routines go through at least two versions, the simple initial one and a later more compact and efficient one.

A strategy such as this could not be implemented without substantial system software purchased at the time the computer was installed. The principal items are shown in Table 2 along with systems software developed by us. Pascal is a recent addition. It combines features of PL/1 and Algol, and although it is quite powerful, we have, as yet, little experience with it.

Our Basic has floating point calculations, including matrix operations. It can call assembly language subroutines and could thus be used for lab instrument data collection and floppy disc storage, for it has file handling capabilities as well. We have used Fortran for this purpose instead because of its greater power and flexibility. The principal reason for having Basic is to provide support to other laboratories. With the software purchases shown in Table 2 we were able immediately to provide both programming and training support. Graduate students and other research people are issued as many of their own discs as they wish. Those who know a little Basic or Fortran are encouraged to work up their data on the minicomputer themselves. We teach them to operate the system and to back up their discs, and we provide extensive, if informal, consulting help. In this way two other laboratories made research use of the machine in its first year. Programs for the analysis of amino acid composition data were prepared by a student in one of these labs. They are included in Table 3.

TABLE 2

SYSTEM SOFTWARE

PURCHASED

    Operating system
    Utilities
        Text editor
        Linkage editor
        Disc file management (copy, delete, pack, list, etc.)
        Miscellaneous others
    An assembler, including macro capability
    Extended basic
    Fortran IV (ANSI compatible)
    Pascal

DEVELOPED LOCALLY

    Cary input driver
    Real time clock handlers
    Tektronix ("soft copy") graphics drivers
    Hewlett Packard ("hard copy") graphics drivers
    Telephone I/O to large remote computers
    Projected
        Cary controller
        High speed paper tape input driver
        Analog instrument input from A/D converter
        Microcomputer I/O drivers

For faculty colleagues who require modest programs, we write them. In addition, as general packages are completed, we make them available along with blank discs and instruction in the use of both machinery and programs. Three laboratories whose people had no prior computer experience whatsoever now use the system in this fashion, and a few of their students have been stimulated to take programming courses as a result. Several other laboratories have expressed interest, and it appears that activity begets activity.

A list of the applications software that is either in use or under development is shown in Table 3. Cary 17 data collection is described in (4) and the general spectroscopic package and molecular graphics are treated in (5). Here, therefore, we describe features of the others that illustrate our "pull yourself up by your bootstraps" approach to program development.

The graphics package was developed to display absorption spectra, the hard copy part having been completed recently. For each graphics device the package consists of a few device dependent assembly routines and a set of short Fortran sub-programs. The latter are sufficiently general that although they were prepared for the Tektronix CRT display, they could be used for the Hewlet Packard (HP) plotter. Device dependent scaling constants had to be changed and a single statement inserted to send a plot initiate signal required by the HP. Calling routines were modified by the insertion of a logical switch to select plot device. Thus, once the HP interface was wired and its assembly language drivers completed, bringing it into research use required only a few hours.

Furthermore, with the availability of hard copy plotting, we discovered an instant demand for keyboard data entry. Column elution profiles of optical density and enzyme activity *vs* fraction number and a great variety of other applications were wanted by nearly every laboratory in the department.

By replacing five or six instructions in each of two Fortran routines, a beginning undergraduate programmer converted spectroscopic plotting into general plotting. The relevant lines had been flagged when the routines were originally written. He then wrote a simple data entry and command loop routine so that a user can format a graph on the Tektronix, which plots at 4800 baud, and then obtain hard copy, which is slower. He did this in two weeks of part time work. His generalized graphics routines will be incorporated into the growing enzyme kinetics library (see Table 3).

Another point worth mentioning is that where we find useful and well documented programs in the literature, we bring them up and are freed thereby of the need to prepare our own descriptive materials.

TABLE 3

APPLICATIONS SOFTWARE

CARY 17D DATA COLLECTION AND RECORDING

GENERAL SPECTROSCOPIC PACKAGE
    Sum and difference curves
    Add signed constant
    Multiply by signed constant
    Light scattering correction
    CRT and hard copy graphics
        Any range of the data, any scaling
        up to five curves
    Multicomponent analysis
        Linear least squares, up to four components
        (stand alone version also for non-spectroscopic
        data entered at the terminal)
    Gauss/Cauchy curve resolution
        Visual curve fit
        Non-linear least squares*

GENERAL DATA PLOT
    Type in and store on disc
    Plot up to five curves
        Data points only or points connected by lines

ENZYME KINETICS
    Direct fit to Michaelis-Menten equation
    Eadie or Lineweaver-Burk plot, line with data
    points*
    Compute K_I for chosen type of inhibition*
    Full time course enzyme assay analysis*
    Complex kinetic equations by non-linear least
    squares*

AMINO ACID COMPOSITION DATA WORK-UP

MINIMUM MOLECULAR WEIGHT

MOLECULAR GRAPHICS*

*Under development

115

Thus the non-linear least squares regression published by Fraser and Suzuki (6) for curve resolution is being readied. The input guesses of the parameters needed by non-linear procedures will be the results of a few easily done attempts at a direct visual fit. Cleland's program (7) for a direct least squares fit to the Michaelis-Menten equation is in regular use. By fitting the rectangular hyperbola directly we avoid the accuracy problems of the Lineweaver-Burk and other similar linear procedures. Even though the user may ultimately wish to display inhibition data in linear forms, his kinetic constants will have been obtained from accurate statistical procedures. The examples chosen here are illustrative. Jennrich and Ralston (8) have provided an excellent general review of data fitting with particular emphasis on obtaining and modifying outside programs to suit one's own computer.

Lastly, it is instructive to consider the economics of a system such as this one. Its purchase cost was $22,000, including the software listed in the top half of Table 2. The hard copy plotter was a gift, and although old and no longer manufactured it is perfectly adaquate. If we allow $3,000 for it, the total system cost comes to $25,000. A conservative estimate of the number of laboratories that will make regular and consistent use of the system is ten, with the total number of laboratories served somewhat higher. The upper limit cost per laboratory is thus $2,500. When the cost of peripheral devices is included in the cost of microcomputer systems, our cost per laboratory is seen to be comparable.

Acknowledgements. The consulting help of Dr. Donald Shombert and his cheerful and enthusiastic encouragement were essential in getting us started. In particular, his help in wiring the computer itself saved us much time and effort, and his expertise in all aspects of computer hardware has often been drawn upon. The aid of the Rutgers University Research Council, a Biomedical Resources Support grant, and the New Jersey State Agricultural Experiment Station are also acknowledged.

Literature Cited

1.   P. Lykos (1975). *Computer Networking and Chemistry*, ed. A.C.S. Symposium Series, No. 19, American Chemical Society, Washington.
2.   P. Lykos (1977). *Minicomputers and Large Scale Computations*, ed. A.C.S. Symposium Series, No. 57, American Chemical Society, Washington.
3.   H. W. Shipton (1979). Ann. Rev. Biophys. Bioeng. *8*, 269.
4.   S. A. Bailey and P. C. Kahn (1980). *Personal Computers in Chemistry*, ed. P. Lykos, Wiley, N.Y. (following paper in this volume).
5.   P. C. Kahn and S. A. Bailey (1980). *Personal Computers in Chemistry*, ed. P. Lykos, Wiley, N.Y. (later paper in this volume).
6.   R. D. B. Fraser and E. Suzuki (1973). *Physical Principles and Techniques of Protein Chemistry, Part C*, ed. S. J. Leach, Academic Press, N.Y., p. 301.
7.   W. W. Cleland (1967). Adv. Enzymol. *29*, 1.
8.   R. I. Jennrich and M. L. Ralston (1979). Ann. Rev. Biophys. Bioeng. *8*, 195.

Minicomputer Data Collection: A Flexible Approach for Multiple Instruments

A general purpose minicomputer system for use in a biochemistry laboratory was described in the previous paper (1), and in the following paper (2) programs for interactive graphics assisted analysis of spectroscopic data are presented. Here we treat the aquisition of that data by the computer from a Cary 17D recording spectrophotometer and the recording of spectra on magnetic disc. The strategy involved in data collection as well as several of the subroutines are quite general and could be used with other instruments or in non-instrumental computing applications.

*Paper of the Journal Series, New Jersey State Agricultural Experiment Station.

[1]Present address: Division of Computer Research and Technology, National Institutes of Health, Bethesda, MD 20205.

Particular emphasis is placed on:
1. the logic of the Cary - computer interface,
2. the program organization of data collection, and
3. real time clock (RTC) routines that allow the taking of readings at precise intervals.

<u>Cary 17D - Computer Interface</u>. The spectrophoto-meter provides digital output through a twenty-six pin connector on its digital volt meter (DVM). Analog to digital conversion was, therefore, not needed. Sixteen of the pins are used to transfer four binary coded decimal (BCD) digits from which the absorbance is constructed. Of the ten pins that remain, five are used to characterize the reading. The other five are redundant. The characterization consists of:
1. data ready status,
2. decimal point location,
3. sign of the reading,
4. 10K bit, and
5. overload status.
These are converted to bit representation, for each has only two states, logical one and zero.
Data ready and sign are self explanatory. The decimal point bit is best illustrated by an example. If the four BCD digits are 1397, the relationship between this bit and the absorbance is:

Decimal point sense line:. 1   0
Format of reading: .XXXX   X.XXX
                        .1397   1.397

The second of these formats arises when the absorbance range of the spectrophotometer is set to two.
The 10K bit is on if the chart range is one or less and if the absorbance is one or more:

10K bit sense line: 1   0
Format of reading: 1.XXXX   0.XXXX or X.XXXO
                      1.1397   0.1397 or 1.397

Note that the 10K bit is never true when the decimal point sense line is false.
The Cary DVM is connected to a general purpose 16-bit digital I/O board, which is mounted in the computer chassis. The four BCD digits are read in parallel, while the five sense lines that characterize the data are input sequentially.
The spectrophotometer displays data in one of

six formats, among which the computer can distinguish
by testing the sense lines. Table 1 shows the logical
values of the sense lines for each of these formats.
There is a unique bit representation of the sense
lines for each form of the reading, and by comparing
the bit map for a given reading with these known bit
representations, one can determine rapidly and simply
the form of that reading.

  Programming Strategy. ANSI standard Fortran was
used wherever possible to simplify the programming.
Assembly language is employed only when it is absolutely necessary. Although our Fortran permits
assembly code to be embedded within Fortran routines,
we restrict assembly operations to separate Fortran
callable sub-programs. Transferability of the Fortran
to other machines is thus preserved. All programs in
either language are thoroughly commented, the number
of comment lines often exceeding the number of lines
of code.
  Each spectral scan is stored on floppy disc as a
separate named file. Data are recorded in binary *via*
unformatted Fortran write statements. Preceeding the
absorbance values in each file is a binary header
record which contains information that describes the
run such as the initial and final wavelengths, scan
rate in nm/sec, eighty characters of title, etc.
  The program begins by asking the user for the
file header, prompting for each item to be entered
into it. The complete header is then displayed and
the user offered the opportunity to make corrections.
When it is satisfactory, it is written to disc.
Figure 1 is a flow chart of this procedure. All
header handling routines are in Fortran.

  Data Collection and Disc Storage. The sampling
time of the Cary DVM is one reading every 200 msec.
Wavelength resolution is, therefore, governed by the
choice of scan rate, which is made at the start of the
run and entered by the user in the header record.
Because wavelength information is not fed to the
computer during the scan, spectra as a function of
wavelength and absorbance at a single wavelength as a
function of time appear identically, and the program
may thus be used also for the recording of enzyme
kinetics data.

TABLE 1

BIT REPRESENTATIONS OF CARY DATA FORMATS

CARY PIN#	FUNCTION	APPEARANCE OF CARY DVM DISPLAY					
		+.XXXX	-.XXXX	+X.XXX	-X.XXX	+1.XXXX	UNSTABLE +1.XXXX
B2	DATA READY	1	1	1	1	1	1
A3	SIGN	0	1	0	1	0	0
B1	DEC. PT.	1	1	0	0	1	1
B3	10K BIT	0	0	0	0	1	1
B4	OVERLOAD	0	0	0	0	0	1
BIT REPRESENTATION		10100	11100	10000	11000	10110	10111

FIGURE 1

CARY SCAN INITIALIZATION

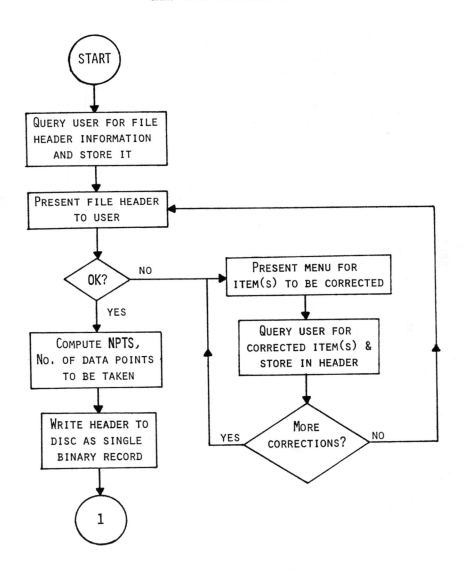

It had been planned that data would be written to the disc as it was obtained, for this would allow the number of data points per run to be limited only by the capacity of the disc ($\approx$ 120,000 values). Tests of the overall time needed to complete a write operation, however, showed that an occasional block transfer took more than 200 msec. Data would thus be lost. All points are, therefore, retained in memory until completion of the run. The present version of the program allows 12,000 absorbance readings per run. Inasmuch as no single scan has yet exceeded 2000 points, the procedure is adequate.

Data output, as mentioned above, is by an unformatted Fortran write statement. The speed of a block transfer could undoubtedly be reduced below the Cary sampling rate if the transfer were done in assembly language, but there is, as yet, no need to do this.

Control over data acquisition is by a Fortran routine. The routine handles timing by calls to assembly language real time clock routines which are described below. It obtains absorbance readings one at a time by a call to an assembly package. Figure 2 is a flow chart of these operations.

The first reading is taken and the clock initialized by setting the time to zero. The program immediately enters a loop all of whose steps are executed well within the 200 msec sampling interval.

Within the loop Fortran calls an assembly language routine to obtain the current time. This it tests to see if 200 msec have elapsed. If not the clock is queried again. When the time limit has passed the clock is reset by a call to another assembly language routine. Fortran then calls the package of assembly language routines that actually take the Cary reading. The steps within the package are as follows:

1. The DATA READY sense line is tested first to see if the spectrophotometer is ready to send a reading. When DATA READY is true the four BCD digits are read in parallel with a single instruction. They are stored as a single 16 bit word.
2. Next the four remaining sense lines are input and the results stored for use in characterizing the reading. To simplify

FIGURE 2

CARY 17 DATA INPUT

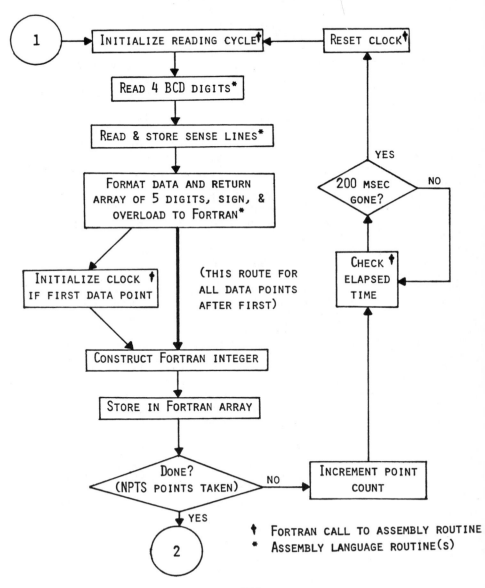

† FORTRAN CALL TO ASSEMBLY ROUTINE
* ASSEMBLY LANGUAGE ROUTINE(S)

124

the characterization the sense line data are compacted into a single computer word, creating a bit map as in Table 1. For example, the logical value of the sign sense line is mapped into bit 1 of the word, that of the decimal point sense line is mapped into bit 2, and so on. Two general purpose assembly language routines perform the necessary operations: BITSET sets a given bit to 0 or 1, and BITEXT can extract the value of a given bit. Comparison with pre-constructed bit masks is also possible.

3. The word containing the BCD digits is then moved by a series of shifts into four words, one per digit.

4. A fifth word is set to zero or one depending upon the bit map established in Table 1. If the absorbance is less than one (first two formats in the table), the extra word is set to zero. It becomes the leading digit in the format 0.XXXX. For the second two forms in the table, the fifth digit is also zero, but it becomes the trailing digit in X.XXX0. If the fifth data format in Table 1 applies, the extra word is set to one, and it becomes the leading digit. A uniform format for all absorbance values of d.dddd is thus obtained.

5. The five digits are returned to Fortran as an integer array of five elements. The logical values of the sign and overload status are also returned.

With control now back in Fortran overload is tested. If true, the absorbance will ultimately be output as the integer value ± 9999 depending on the value of the sign. If this is to be the case further processing of the digits is not needed, and it is bypassed.

If the value is within the spectrophotometer limits, the array of five digits is converted to a single integer whose value is the absorbance times 10,000. The integer value is stored in an array, the point counter tested to see if the scan is completed, and, if not, the loop is repeated.

Data transfer to the disc is flow charted in Figure 3. It is performed in Fortran, and after it is complete a message is issued to the user, control over the real time clock is restored to the operating system by a call to an assembly language subroutine, and the program terminates.

The storage of data as integers whose values are the absorbance times 10,000 has the advantage of economy in both memory and disc usage. Since floating point data require double the space, two words per point instead of one, there would be room in memory for 6000 rather than 12,000 values if floating point were to be used. Integer storage has the disadvantage, however, of making it impossible to store absorbance values greater than 3.2767. When multiplied by 10,000, this would yield the largest signed integer that can be represented in a single sixteen bit word. For our purposes, however, this is not a problem, for the dynamic range of the Cary DVM, -0.1049 to +1.9999, remains well within the sixteen bit limit after conversion to integer. The averaging of multiple scans as the data are taken would have to be coded carefully to avoid integer overflow, but this is not an insurmountable problem.

Post-experiment signal averaging *is* possible with the present procedures, in which each scan is collected as a separate disc file. The averaging could be carried out interactively as described in the following paper (2). This is adequate provided that the number of scans to be averaged is not too large ($< \simeq 10$), and it has, in fact been applied to obtain standard spectra to be used in least squares composition analysis of multicomponent mixtures. For more than approximately ten scans, however, interactive averaging becomes inconvenient.

Until the spectrophotometer's wavelength drive is placed under computer control, though, the taking of data for many scans will not be practical in any case. Once computer control - as opposed to passive data acquisition - is implemented, computation and/or I/O after each scan could be programmed without difficulty.

Since no more than 2000 data points have been recorded in a single spectrum, allowance for 12,000 might seem to be a luxury. Programs for analyzing

FIGURE 3

CARY DATA OUTPUT TO DISC

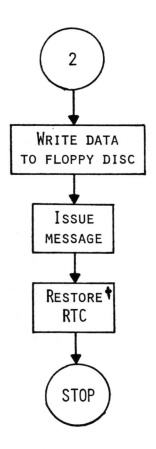

✠ FORTRAN CALL TO ASSEMBLY ROUTINE

the full time course of spectrophotometric enzyme assays, however, are presently under development. In these experiments data are taken as a function of time at a single wavelength. From one or a few such runs one can obtain in some cases data that would require many initial velocity assays and in others kinetic constants that are not otherwise obtainable at all (3 - 6). Some experiments of this kind will require the full 12,000 data points that are presently available.

Real Time Clock Routines. An integral part of the package is the set of routines that utilize the computer's real time clock. There are five short assembly language programs, all of which are callable from Fortran.

The clock generates an interrupt signal every 10 milliseconds, and a clock "tick" thus represents 10 msec of elapsed time. The timing interval is set by a jumper on the processor board and could be set to 1 msec or 100 msec if needed. When the clock sends an interrupt pulse, program control goes directly to a specific location in memory to execute the code stored there. The operating system normally uses the clock to maintain the time of day, the month, and the year. The start routine, which is called when data collection begins (see Figure 2), short circuits the operating system's extravagant time-keeping code by inserting at the interrupt location a simple counter that is incremented each time the clock ticks. When data collection is complete, another routine restores control of the clock to the operating system, for the system crashes if this is not done.

The three other routines in the package are used to perform the basic tasks that are done for any clock. One will reset the clock to zero so that timing can start from a known point in a program. Another returns the elapsed time since the last reset, and the last lets the program wait for a specified length of time as passed in the call statement.

Summary. It is intended that data be collected from other instruments, so the programs are as modular and general as possible. Data interpretation procedures will be specific for each instrument,

of course, but several utility routines and the real
time clock programs can be used with any instrument,
as can the Fortran control logic. By programming in
Fortran whenever possible, moreover, the modification
of existing programs and the development of new
applications that use existing subroutines is simplified. A price is paid in the efficiency of program
execution and, to some extent, of memory usage.
Except in tasks requiring the most stringent timing,
however, these drawbacks are outweighed by the
advantages.

Literature Cited

1.   P. C. Kahn (1980). *Personal Computers in
     Chemistry*, ed. P. Lykos, Wiley, N.Y.
     (preceeding paper in this volume).
2.   P. C. Kahn and S. A. Bailey (1980). *Personal
     Computers in Chemistry*, ed. P. Lykos, Wiley, N.Y.
     (following paper in this volume).
3.   C. Frieden (1975). J. Biol. Chem. *250*,
     2111-2113.
     C. Frieden and J. Fernandez-Sousa (1975). J.
     Biol. Chem. *250*, 2106-2111.
4.   S.-L. Yun and C. H. Suelter (1977). Biochim.
     Biophys. Acta *480*, 1-13.
5.   S. A. Bizzolero, A. W. Kaiser and H. Dutler
     (1973). Eur. J. Biochem. *33*, 292-300.
6.   I. G. Darvey, R. Schrager and L. D. Kohn (1975).
     J. Biol. Chem. *250*, 4696-4701.

P. C. KAHN and S. A. BAILEY

CHAPTER 11

Interactive Graphic Assisted Analysis
of Spectroscopic Data

     The selection and development of a general pur-
pose minicomputer for use in the biochemistry
laboratory was described in the first paper in this
set (1) and the online collection and storage on disc
of absorption spectra in the second (2). Here we
present an interactive program for graphics assisted
analysis of spectroscopic data. The software design
is sufficiently general that it could be applied
easily to applications other than spectroscopy. One
could, in fact, define one's own interactive language

*Paper of the Journal Series, New Jersey State Agri-
cultural Experiment Station.

[1]Present address: Division of Computer Research and
Technology, National Institutes of Health, Bethesda,
MD 20205.

for any purpose whatsoever, implementing it in any
language on either a minicomputer, as here, or on a
smaller machine. Available memory is the only restriction on the size and complexity of one's interactive language, and this restriction can be avoided
altogether if program overlay facilities are available.
    The functions performed by the present package
are listed in Table 1. Most are self explanatory and

TABLE 1

PRINCIPAL FEATURES OF
INTERACTIVE SPECTROSCOPIC PROGRAM

---

1.   Up to five curves of 1700 data points each can
      be handled at one time

2.   Construct sum and difference curves

3.   Multiply by or add signed constant

4.   Light scattering correction, method of Leach and
      Scheraga, ref. 11

5.   CRT and hard copy graphics
          Any range of the data, any scaling. Up to five
          curves per CRT plot. No limit on hard copy

6.   Composition analysis of multicomponent mixtures.
          Linear least squares, up to four components

7.   Gauss/Cauchy curve resolution
          Visual curve fit
          Iterative non-linear least squares*

---

*Under development

---

represent the standard operations which one would want
to carry out on absorption spectra. Before detailing
the logic of their implementation, however, it is
useful to present a sample of the interactive language.
This we do by illustrating the user's commands and

system responses for a working session. The principal
command language subroutines are then described,
following which a typical problem in curve resolution
is examined. Lastly, a different but related package
for the display of molecular structure is outlined
and our application of it treated briefly.

A Working Session. The beginning of a session
is shown in Figure 1. The program prompts the user
with COMMAND? This prompt would appear between each
line of the figure and throughout the session, but we
have omitted it here for clarity. The user responds
to the prompt with a single line which contains a
command word followed by one or more parameters. The
parameters may be numerical or alphabetic depending
on the command. The prompt and the user's command
lines are echoed onto the line printer, as are the
program's responses to the commands. This yields a
complete printed record of the session, which is useful in tracking down errors and in compiling results
for publication, etc.
    Command input is normally from Fortran logical
unit five. This is usually assigned to the key board,
but it can be assigned to the disc, allowing a file of
previously stored commands to drive the program in
unattended batch mode. A command exists, also, to
change the logical unit for command input, which permits mixed batch and interactive operations in a
single session. Repetitive processing of several
curves could be done this way, for example, or a
frequently used group of commands saved for convenience.
    Having constructed in Figure 1 baseline sub-
stracted curves one and two, the user examines them
on the Tektronix graphics terminal's screen with the
result shown in Figure 2. Although the spectra of
these samples was recorded at 0.2 nm intervals, the
plot in Figure 2 and in subsequent figures is at 1 nm
intervals. The user may change the plot interval at
will. Calculations, however, are done on all data
points.
    The session continues in Figure 3, where the user
narrows the wavelength range, expands the absorbance
scale to fill the plot, and obtains the hard copy
shown in Figure 4. The quality of these hard copy

FIGURE 1

BEGINNING OF INTERACTIVE WORKING SESSION

COMMAND?

, SYSTEM PROMPT.   SEE TEXT.

SREAD   5   1

, READ CURVE 5 FROM FORTRAN UNIT 1.   FIVE CURVES OF
, UP TO 1700 DATA POINTS EACH CAN BE HANDLED IN
, MEMORY.
,
, ALL COMMAND LINES ARE PRINTED TO GIVE USER A RECORD
, OF THE SESSION,   A BLANK FOLLOWED BY A COMMA CAUSES
, THE LINE TO BE PRINTED AS A COMMENT.
,
, NOTE FREE FORMAT FOR INPUT.

SREAD   4   2

SUB 1   4   5

, LET CURVE 1 = CURVE 4 - CURVE 5.
, THIS COULD BE A BASELINE SUBTRACTION OR A
, DIFFERENCE CURVE.

SREAD   2   3

SUB 2   2   5

, LET CURVE 2 = CURVE 2 - CURVE 5
, CALCULATED CURVE REPLACES ORIGINAL

TRANGE 450   720   0 1

, SET WAVELENGTH AND ABSORBANCE RANGES FOR PLOT

PLOT 1   2

, FIGURE 2 IS A PHOTOGRAPH OF THE TEKTRONIX SCREEN.

133

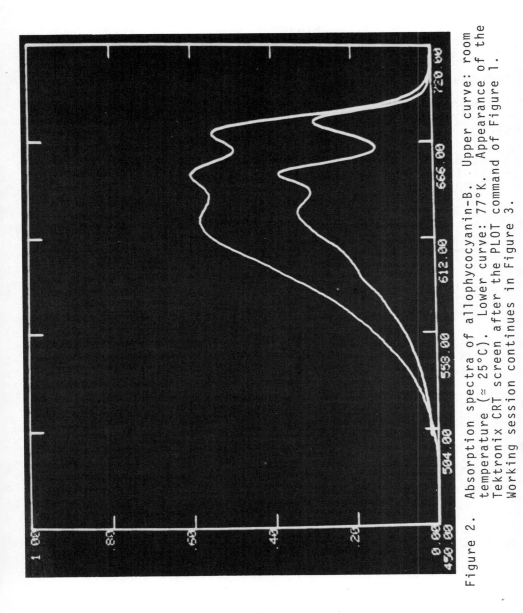

Figure 2.   Absorption spectra of allophycocyanin-B. Upper curve: room
temperature ($\approx$ 25°C). Lower curve: 77°K. Appearance of the
Tektronix CRT screen after the PLOT command of Figure 1.
Working session continues in Figure 3.

FIGURE 3

WORKING SESSION - CONTINUED

, RANGE NO GOOD. TRY AGAIN.

TRANGE 500   700 0 0.8

, GET HARD COPY

PLTDEV 1

, (ZERO SELECTS THE TEKTRONIX, WHICH IS THE
, DEFAULT VALUE)

PLOT 1 2

, THE PLOT IS SHOWN IN FIGURE 4.

---

plots is good enough that they can be included
directly in a manuscript after hand lettering the co-
ordinate axes.

In Figure 5 the user defines a set of Gaussian
bands, constructs their sum, and obtains a plot of the
individual bands, the experimental curve, and the
fitted curve (Figure 6). Since this is a simulated
session offered to illustrate the use of the interactive language, the final fit is shown in Figure 6.
One's first guess at a fit, of course, would not be
as good. By re-defining one or more band parameters
the user can refine the fit until it is satisfactory.
A PRINT BANDS command (not shown) would then produce
a table of the currently defined bands with their
parameters along with the areas under each band in
c.g.s. dipole strength (Debye) units. The present
version of the curve resolution routines allow up to
twenty bands to be defined. The CONSTR (Figure 5)
command, however, constructs the sum only of the bands
given in the command, so that the bands resolving two
or more curves can be retained simultaneously in
memory. This is of particular use in the resolution
of circular dichroism and absorption spectra of the
same sample, wherein bands of the same position and
halfwidth occur in the two observables, the spectra

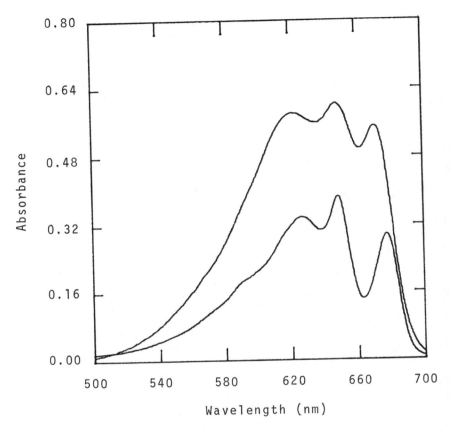

Figure 4.   Spectra of Figure 2 after re-scaling axes in Figure 3.   Hard copy plot.

---

differing only in the signs and peak intensities of the transitions (3,4).

The bands in the example shown here are pure Gaussians. Mixed Gaussian/Cauchy functions may be used, with the fraction of each component varied by the user at will between zero and one. Whatever fraction is chosen, however, is applied to all bands in a resolution.

The curve resolution procedure described here mimics the operation of the Dupont 310 Curve Resolver,

FIGURE 5

WORKING SESSION - CONCLUDED

, GAUSSIAN CURVE RESOLUTION

, FIRST DEFINE A SET OF BANDS:

BAND 1   673.5   0.48   11.7

, BAND #1 IS AT 673.5 NM, 0.48 ABSORBANCE AT PEAK,
, 11.7 NM HALFWIDTH

BAND 2   652.2   0.408   14.5

BAND 3   624.5   0.54   25

BAND 4   593.2   0.252   21.8

BAND 5   562   0.14   25

CONSTR 3 1 2 3 4 5

, CONSTRUCT CURVE 3 AS THE SUM OF GAUSSIAN BANDS 1
, THROUGH 5.

BDRAW ALL 2 3

, DRAW ALL BANDS PLUS CURVES 2 (EXPT'L) AND 3
, (CALCULATED)

, THE HARD COPY PLOT IS SHOWN IN FIGURE 6

---

an analog device which is no longer made. Using a
half silvered mirror, the Dupont instrument would
superimpose one's experimental data on the current
sum of bands, and by varying one control for each band
parameter, one could adjust the fit as desired.

   Defining An Interactive Language. The interactive procedures described here are implemented
entirely in Fortran. The logic is sufficiently simple,
though, that it could be implemented easily in any
other language, including Basic or assembly language.

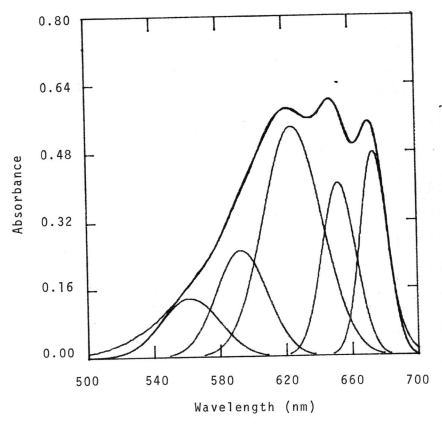

Figure 6.   Gaussian curve resolution of allophyco-
cyanin-β room temperature spectrum. Five
band fit produced by commands of Figure 5.

It originated (5,6) in Levinthal's laboratory, where
it is used in PAKGGRAF, a Fortran program for the
simulation and manipulation of protein structures. Our
version is a modified and simplified form of the
PAKGGRAF command logic.
     The present package contains approximately fifty
subroutines, most of which perform the functions in
Table 1. With appropriate main programs these could
just as well be used in stand-alone, non-interactive

processing.   Interactive operations require only four
subroutines.   These, along with brief synopses of
their functions, are listed in Table 2.   The package

TABLE 2

PRINCIPAL SUBROUTINES

(FORTRAN)

RDLINE

Reads a command line, parses it, and provides
command parameters to processor *via* common.

COMMAND PROCESSOR(S) - COMNDn ROUTINES

Processes command parameters and dispatches to
appropriate subroutines.

SREAL

Tests numerical command parameters in alphabetic
format to be sure that only valid ASCII numerals,
+, -, and E are present.   Returns floating point
value of parameter (even if actual value is
integer - see text).

TESTn ROUTINES

Examine integer or floating point parameters after
SREAL and other conversions to be sure that they
are within legal bounds.   See text.

---

is thus divided rigidly into two logical parts.   The
first acquires and interprets commands, checks that
they and their parameters are valid, and dispatches
to the appropriate processing routines, which, taken
together, constitute the second part.   Processing
routines, in turn, do no error checking of their
arguments or program commands, which makes them more
compact than would otherwise be the case.

At the heart of the interactive logic is a pair of arrays, COMNDS, each element of which contains a single command word in ASCII representation, and PARAMS, which contains all the parameters of the user's current command line. They are shown in Figure 7. In the present implementation we allow up to eight characters per parameter, which is the length of a double precision Fortran variable on our sixteen bit machine. This could be reduced to four or even two characters if memory were to be limiting. In making such a reduction, however, one would have to take into account two factors:

    1.   The first parameter of a command line, since it is the command word itself, must have the same length as the elements of COMNDS or the test at line 12 of Figure 7 will fail.

    2.   Provision must be made to input floating point or fixed decimal data in parameters. These generally exceed four characters in length, e.g. -1.32E-5, 462.3.

Array PARAMS is placed in a Fortran common to make it available to a variety of routines: RDLINE fills it, and COMND1 and other dispatching routines examine it. Also, in PARAMS common are MAXPRM, an integer initialized to the dimension of the array and used in tests to prevent writing into or loading from memory beyond the array bounds, and NPARM, which is set by RDLINE to the number of parameters in the current command line. NPARM is also used in tests.

    <u>Command Line Input and Decomposition</u>. Subroutine RDLINE, called in statement 900 of Figure 7, receives the prompt to be issued to the user as its first argument (see Figure 1, top line). The prompt is limited to 72 bytes, the length of a teletype line. INUNIT is the Fortran logical unit number from which the command line is to be read. As noted above, a command exists to change this. NBYTES is an upper limit for the length of the command line. It, too, is generally set to 72 characters. If it is zero, however, RDLINE returns immediately after issuing the prompt, thus providing for a convenient way to issue a one line message without coding any I/O or format statements. Memory is thus conserved, and, looking toward the eventual use of program overlay, I/O

FIGURE 7

COMMAND PROCESSOR ARRAYS & LOGIC

(Simulated Fortran Code -
For Illustrative Purposes Only)

```
      SUBROUTINE COMND1                                    1
C VARIABLE DECLARATIONS BEGIN HERE                         2
      DOUBLE PRECISION COMNDS(10), PARAMS(20)              3
      COMMON/PARAMS/MAXPRM, NPARM, PARAMS                  4
      DATA MAXPRM/20/                                      5
      DATA COMNDS/'SREAD   ','PRINT   ','ADD     ',        6
     1 'SUB     ','MULT    ','PLOT    ','PLTDEV  ',        7
     2 'BDRAW   ','BAND    ','        ','/,NACTS/10/       8
C EXECUTION BEGINS WITH ENTRY TO MAIN COMMAND LOOP         9
C AT 900
      900 CALL RDLINE('COMMAND?', INUNIT, NBYTES)         10
      DO 1 I = 1, NACTS                                   11
      IF (PARAMS(1) .EQ. COMNDS(I)) GO TO 3               12
    1 CONTINUE                                            13
C IF HERE, COMMAND NOT FOUND. LOOK ELSEWHERE.            14
      CALL COMND2                                         15
C NO RETURN HERE. COMND1 CALLED INSTEAD TO RE-ENTER      16
C LOOP AT 900
C NEXT STATEMENT BRANCHES TO COMMAND WORD                17
C PROCESSING CODE.
    3 GO TO (101, 201, ... 1001), I                       18
C FIRST COMMAND IS 'SREAD' - READ A SPECTRUM.            19
C SEE FIG. 1.
  101 ISPEC = SREAL(PARAMS(2))                            20
      IUNIT = SREAL(PARAMS(3))                            21
      CALL TEST1(2, ISPEC, 0.0)                           22
      CALL TEST1(4, IUNIT, 0.0)                           23
      CALL SREAD(ISPEC,IUNIT)                             24
      GO TO 900                                           25
C ONLY DISPATCHING ROUTINES CLOSE THE LOOP BY            26
C CALLS TO COMND1.
C COMMAND EXECUTION ROUTINES, IN ORDER TO BE USED        27
C IN OTHER
C PROGRAMS, HAVE NORMAL FORTRAN RETURNS.                 28
C                                                        29
C SECOND COMMAND - PRINT SOMETHING.  CALL PRINT          30
C DISPATCHER
  201 CALL SPRNT1                                         31
```

(Continued)

141

FIGURE 7 (CONTINUED)

```
C THIS, TOO, CALLS COMND1 TO CLOSE THE LOOP          32
C                                                    33
C THIRD COMMAND - ADD TWO SPECTRA TOGETHER TO        34
C MAKE A SUMMED CURVE
  301 ..........                                     35
```

statements can be restricted to a few routines.
RDLINE fills array PARAMS with the ASCII characters of the user's command line. The first parameter
is left adjusted in its field of eight characters with
blanks padding on the right as needed. This sets up
PARAMS (1) for comparison with the list of allowed
commands at line 12, Figure 7. Parameters after the
first are right adjusted by RDLINE so that later
numerical conversions in Fortran do not add spurious
trailing zeroes where blanks are encountered. RDLINE
searches for either a comma or a blank as a parameter
terminator, taking any subsequent non-blank character
as a parameter beginning. This provides for free
format in command input. RDLINE also checks to be
sure that the maximum parameter length, eight
characters, is not exceeded.

Command Word Analysis - COMNDn Routines. Subroutine COMND1 begins execution with the main command
loop at the call to RDLINE. It is therefore the
effective main program, the actual main being a dummy
that simply calls COMND1. The command loop is very
short, consisting of the call to RDLINE plus the next
five executable statements. PARAMS (1) is tested
against each of the elements of array COMNDS by the DO
loop at lines 11 through 13 of Figure 7. If the user's
command is found, the GO TO at line 12 exits the DO
loop with the index number in array COMNDS of the
found command. If the user's command is not found,
the DO loop is completed, and COMND2 is called.
COMND2 is identical in structure to COMND1 except that
it processes commands not found in COMND1. The CONSTR
command in Figure 5 would be an example. If the
user's request is not found there, COMND3 would be
called. The last COMNDn subroutine in the chain is
reached only if the user has made an error and entered
an invalid command, and this subroutine therefore calls

an error message issuing routine and then calls
COMND1 to close the main command loop. The interactive language is thus easily expanded by adding
new COMNDn routines.

If program overlay were to be used, the COMNDn
routines could replace one another, and once the
user's command is identified, the processing subroutines for that command could be brought into memory,
replacing those used for the previous command. We
have not yet had to use overlay facilities, however,
for our present language has only about thirty
commands. With one exception all routines fit into
our 32K words of memory. The logic was developed with
overlay in mind, though.

The exception is that the routines for multicomponent composition analysis and curve resolution
cannot be resident in memory at the same time. To
accomodate them, two forms of the package have been
link edited. In one the curve resolution section is
present but the composition analysis part is represented by a dummy routine that does an immediate
return. In the other the reverse is the case. The
command processing is the same in the two versions.
In both forms the data arrays and program commons are
all at the very beginning of program memory, and they
occur at exactly the same addresses. This has the
consequence of allowing us to fill the data and common
areas with either version, exit that version, and load
the other, which can then use the previously filled
data section.

One could call this the "poor man's overlay". We
describe it because it would allow laboratories
whose machines do not have pre-programmed overlay
facilities to construct, in effect, their own.

   Command Parameter Analysis and Dispatching. When
the user's command is found and the DO loop exited,
the DO index, I, is used in statement 3 as the argument
of the computed GO TO. That statement thus transfers
control to a block of code specific to the I'th
command in array COMNDS. What happens next depends
upon the particular command, there being two classes
of response. The first is further command analysis,
with the call to SPRNT1 an example. SPRNT1, organized

identically to COMND1*, examines the *second* of the
user's command parameters (PARAMS(2)) in its DO loop
to determine what is to be printed. If its set of
possibilities is exhausted, it calls SPRNT2, etc., in
a chain similar to the chain of COMNDn routines. The
PRINT BANDS request, mentioned earlier, is processed
in this way. Other print operations include all or
part of the absorbance data of one or more spectra in
tabular format, file headers, the current values of
various program options, etc.
     The second class of response, in which the calls
to command execution subroutines are set up, is
illustrated by lines 20 through 25 in Figure 7. The
command being processed is SREAD, examples of which
are shown in Figure 1. Function SREAL receives as its
argument the user's numerical parameters. If these
contain only the digits zero through nine, the plus
or minus sign, the letter *E*, and blanks, the ASCII
parameter is converted to a Fortran single precision
floating point variable, which is returned. This
routine exists to avoid the abnormal program termi-
nation that a user typing error would produce, for an
invalid character in a numeric field is detected by
Fortran, which cannot handle it. With SREAL, a user
error leads to a "soft return", i.e. a call to the
error message routine followed by a call to COMND1
for another try.
     The assignment statements at lines 22 and 23 in
Figure 7 force conversion of the real value returned
by SREAL to integer, ISPEC for the first SREAD in
Figure 1 being five and IUNIT one.
     Before ISPEC and IUNIT are passed to subroutine
SREAD, which reads the desired spectrum from disc, the
parameters are examined to be sure that they are within allowed bounds. ISPEC, for example, as the curve
number within memory, must fall between one and five
or the read-in will place the data outside the data
arrays, leading to immediate disaster. TESTn routines
perform tests of this kind. These subprograms receive
as arguments the parameters to be tested and a code,
IOP, which is the first argument in the calls to
TEST1 in Figure 7. IOP is used in a computed GO TO

*The names of the print operations in the SPRNT1 COMNDS
array are right adjusted to coincide with the specifications of RDLINE discussed above.

analogous to that of COMND1 to branch to the appropriate block of code. Since both integer and real
arguments may have to be tested, TEST1 allows arguments
for each. The section of code selected by IOP governs
the argument that is used. For complete generality a
double precision argument could be added.

Should the test indicated by IOP not be found in
TEST1, TEST2 could be called in a manner similar to
the COMNDn chain. We have not yet had to do this,
however, as the tests presently in use still fit
within a single routine of moderate size.

Some Comments on Efficiency. Upon first examination it might seem that converting the user's
integer parameter to floating point in subroutine
SREAL and then back again to integer is wasteful and
inefficient. Direct integer conversion of the ASCII
parameter would certainly take less time, but it would
require us to write the code to do it. Since we had
to have floating point code in any case, the present
procedure uses the Fortran run-time library for the
integer conversion. There are thus savings both in
development time and in memory usage.

These considerations do raise the general question
of efficiency, however. In preparing software,
especially on small machines, one must define clearly
and in advance the form of efficiency that is to be
sought. For the processing of interactive commands
speed of execution is not a serious constraint, for the
relevant time scale is human response time. Although
SREAL and associated conversions are slow by computer
standards, the response to the user is virtually
instantaneous. The computed GO TO in command word
identification is another place where the well known
computational inefficiency of the Fortran statement
is not a drawback, and its simplicity is an asset.

Subroutines that are primarily computational, such
as those involved in curve resolution, have quite
different requirements. These we optimize for speed.
Since our "integer only" hardware does its floating
point calculations by software, this is particularly
important.

A Problem in Curve Resolution. The spectra shown
in this paper are of allophycocyanin-B, a protein
component of the *Nostoc* sp. photosynthetic apparatus.

The protein was purified by Greenwald and Zilinskas (7). The upper curve in Figures 2 and 4 was recorded at room temperature and the lower at 77°K.

In the resolution of the room temperature spectrum shown in Figure 6, five Gaussian bands produce an excellent fit from 550 nm to 690 nm. The data outside this region were not of interest, and no attempt was made to fit them.

The narrowing of the bands at liquid nitrogen temperature, however, leads to better inherent resolution, as is seen in Figure 4. Since the electronic transitions present at 77°K are also present at room temperature and occur with rare exceptions (8) at approximately the same wavelengths, curve resolution of the low temperature spectrum should provide a more reliable indication of the minimum number of transitions present. Resolution of the 77°K spectrum does, in fact, require an additional band at 668.6 nm to fit the deep trough in that vicinity. This is shown in Figure 8.

When six bands are used to resolve the room temperature spectrum, however, an ambiguous result is obtained. Based upon simple visual inspection the resolutions of Figures 9 and 10 both fit the data. Curve resolution, to be sure, does not always provide a unique set of bands (4, 9), and one must, therefore, be cautious in interpretation. In the absence of outside information the principle of Occam's Razor applies: use the minimum number of bands. The match between experimental and calculated curves can be made as good as one wishes by adding more bands, but the result would make little spectroscopic sense. Here, however, outside information is supplied by the presence of the sixth band in the 77°K spectrum.

One approach to removing the ambiguity would be to subject the results to statistical tests. To this end an iterative non-linear least squares program is being developed. Another approach would be to resolve in parallel with the absorption data another experimentally observable property of the same electronic transitions such as the circular dichroism spectrum. Additional constraints on the fitting are thereby imposed. For our purposes here, however, the essential point is that interactive graphics assisted procedures are ideally suited to the rapid identification of such problems and to the application of chemical insight in

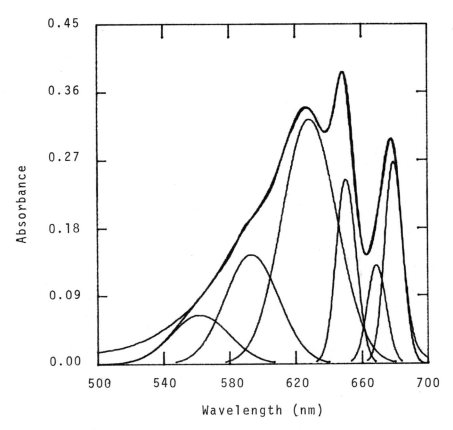

Figure 8.   Gaussian curve resolution of allophyco-
            cyanin-B spectrum at 77°K. Six band fit.

---

their solution.

   The Display of Molecular Structures. The stereoscopic images in Figure 11 were produced on the
present system. They represent a section of the α
chains of hemoglobin as described in the figure legend.
A hydrogen bond exists between the imidazole ring of
a histidine side chain, whose hydrogen atom projects
upward in the center of the picture, and a carbonyl
group of the polypeptide backbone, whose oxygen

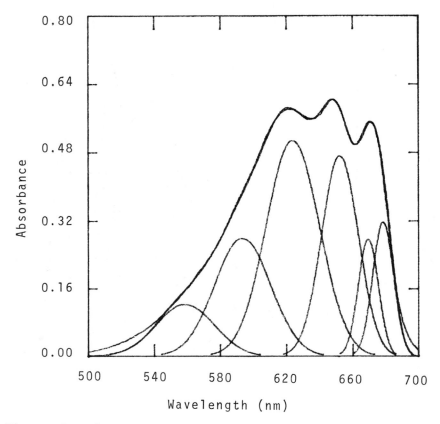

Figure 9.   Gaussian curve resolution of allophycocyanin-B room temperature spectrum. Six
band fit keeping wavelengths close to
those of Figure 8.

projects downward. It is seen that the H-bond is
nearly linear in the oxy conformation and distinctly
non-linear in the deoxy form. The effect of this
difference in geometry on the strength of the H-bond
and on functional properties of the protein has been
discussed by Valentine, *et al* (10), from which Figure
11 is taken. The difference was first noticed in
studying graphical images.

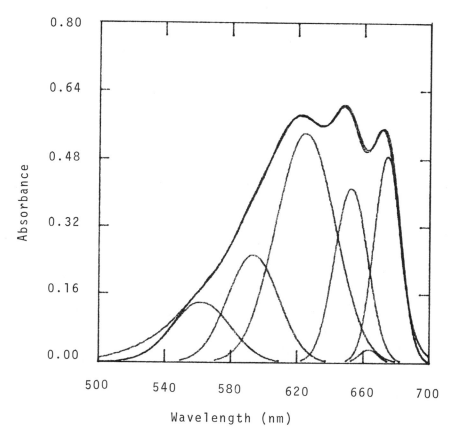

Figure 10.   Gaussian curve resolution of allophycocyanin-B room temperature spectrum.
Alternate six band fit (see text).

The display of molecular structures is not new.
With the increasing availability of highly refined
structures, however, the possibility for detailed
structure-function correlation grows.   There are, in
addition, obvious applications to teaching at all
levels.   Whatever one's interests, one requires the
ability to alter the display at will, adding or
removing parts of the molecule and rotating it in
order to focus attention on the relevant structural

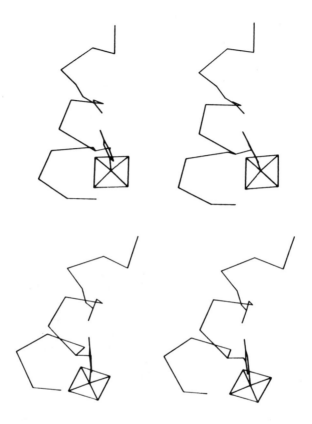

Figure 11. Stereoscopic images of axial imidazole
hydrogen bonds in the α chains of horse
deoxy- (upper) and methemoglobin (lower).
Methemeglobin has the oxy- conformation.
Section of F helix from Ala 79 through Arg
92. For Leu 83 all nonhydrogen backbone
atoms are drawn. The peptide carbonyl of
this residue is the hydrogen bond acceptor.
For all other residues only α carbons are
drawn. The imidazole is seen edge on to
the right of the helix. The heme group is
represented by the iron atom and by a
square connecting the four pyrrole nitro-
gens. From reference 10.

features. Figure 11 contains only a few of the several
thousand atoms in the protein.
 Interactive programs similar in structure to the
one described here are well suited to the purpose.
PAKGGRAF (5, 6), whose command logic we have adapted,
is built around molecular display capabilities of this
kind.
 Storage of atomic coordinates and associated data
on magnetic media, moreover, allows molecules of any
size to be displayed, for the picture can be constructed "on the fly" without retaining in memory
data for many atoms. Microcomputers of limited memory
could thus be used. The loading of coordinates from
large machines over a telephone line onto local
magnetic devices, in addition, makes available an
enormous library of data. Our interests are in the
spectroscopy of proteins (4) for which the necessary
calculations can only be done on a large machine. They
are much too expensive to do on a time share basis,
however, which has led to our commitment to the
present minicomputer system.

Literature Cited

1.   P. C. Kahn (1980). *Personal Computers in*
     Chemistry, ed. P. Lykos, Wiley, N.Y.
     (earlier paper in this volume).
2.   S. A. Bailey and P. C. Kahn (1980). *Personal*
     Computers in Chemistry, ed. P. Lykos, Wiley, N.Y.
     (preceeding paper in this volume).
3.   A. Moscowitz (1960). "*Optical Rotary Dispersion*",
     (C. Djerassi, ed.) McGraw-Hill, N.Y. p. 269.
4.   P. C. Kahn (1979). Meth. Enzymol. *61*, 339.
5.   L. Katz and C. Levinthal (1972). Ann. Rev.
     Biophys. Bioeng. *1*, 465.
6.   C. Levinthal, personal communication.
7.   L. Greenwald and B. A. Zilinskas (1979).
     Submitted for publication.
8.   B. A. Zilinskas, L. Greenwald, C. Bailey and
     P. C. Kahn. Biochim. Biophys. Acta, in press.
9.   I. Tinoco and C. R. Cantor (1970). Meth. Biochem.
     Anal. *18*, 81.
10. J. S. Valentine, R. P. Sheridan, L. C. Allen and
     P. C. Kahn (1979). Proc. Nat. Acad. Sci. USA
     76, 1009.

11.   S. J. Leach and H. A. Scheraga (1960).   J. Amer. Chem. Soc. *82*, 4790.

Numerical Methods on an Interactive
Graphic Microcomputer

INTRODUCTION

The Tektronix 4051 and its newer siblings the 4052 and 4054 are intelligent, high resolution graphic micro-computers with their 32K ROM BASIC interpreters and featuring 14 decimal digit accuracy. This combination of arithmetic accuracy, with 1024x780 addressable points on an 8x6 inch storage type CRT, makes possible the early introduction of numerical techniques into the undergraduate chemistry curriculum. Although numerical and graphic techniques have become common to most engineering curricula since the advent of computers, their use in the liberal arts and sciences has been limited by the lack of low cost easy to use graphic hardware and appropriate software. The stand alone nature of the 4051 (with up to 32K of RAM) with its hard-wired BASIC interpreter and 300 K tape storage facilities makes possible the building of a library of numerical procedures which can be readily accessed by most users with a minimum of difficulty. Tektronix Inc. offers a series of graphic tapes in engineering, mathematics, and statistics for approximately $400 per tape which address themselves to many applied numerical problems. However, some of their programs are secret; so, they cannot be listed or copied. On the other hand, the convenient graphic command structure makes it relatively easy for the user to write his own graphic programs at virtually any desired level of sophistication. This latter feature is

especially valuable from a pedagogical viewpoint, and
it is the purpose of this paper to describe briefly
the utilization of a variety of these numerical techniques in physical chemistry.

APPLICATIONS

For many problems there is an exponential relationship between the dependent and independent vari-
able(s) as for example in the case of the temperature
dependence of the equilibrium constant, vapor pressure as a function of temperature, and a variety of
rate processes. In the case of vapor pressure measurements, most often the student is introduced to the
Clausius-Clapeyron equation in its logarithmic form
which can be readily fitted by linear regression or
least squares techniques in which the deviations in
the logarithms of the dependent variable are minimized with respect to the coefficient $C(1)$ and $C(2)$
in equation (1) below. It is also shown that this
process is equivalent to minimizing the relative deviations in the vapor pressures, since for small but
finite deviations, $\Delta P/P$ is approximately equal to
$\Delta(\ln P)$.

Inclusion of the temperature dependence of the
heat of vaporization in the Clapeyron equation and
subsequent integration results in the so-called modified Kirchoff equation below:[1]

$$\ln P = C(1) + \frac{C(2)}{T} + C(3) \ln T + C(4) T \qquad (1)$$

which is no longer a linear function of $1/T$. Furthermore, since the pressures are usually measured with a
mercury manometer having a fixed absolute error, it
is the deviations in the exponential form of equation
(1) which should be minimized in determining the coefficients. If the pressures are measured with a
gauge having a fixed relative error, equation (1) can
be handled by the usual least squares procedure since
the normal equations are linear in the coefficients.
In either case the data fitting procedure must be consistent with the experimental errors involved as noted above. To this end a handout is provided with
appropriate references showing how equation (1) or

its exponential form can be handled by a Taylor series approximation about an initial set (Clausius-Clapeyron) of values of the parameters and which is linear in the correction terms dC(1), dC(2), dC(3), and dC(4) in an extension of the procedure described elsewhere.[2] The results of such a program written by the author are shown in Figure 1 and 2 in which the slight departure from linearity is readily observed. The data set for trimethylamine is taken from the very precise work reported by Aston.[3]

FIGURE 1

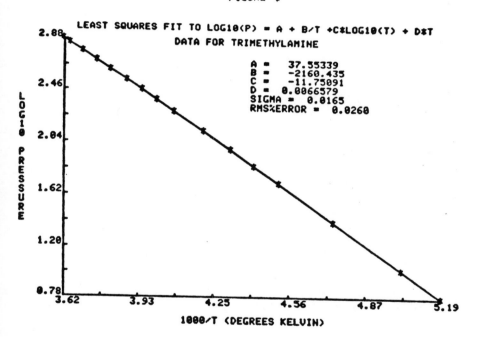

Figure 1 illustrates the results of minimizing the absolute deviations as noted by the minimum in the standard deviation in the pressures. Figure 2 was obtained by minimizing the relative deviations, using the approximation noted above, and as indicated by a minimum in the root-mean-square (RMS) percentage error.

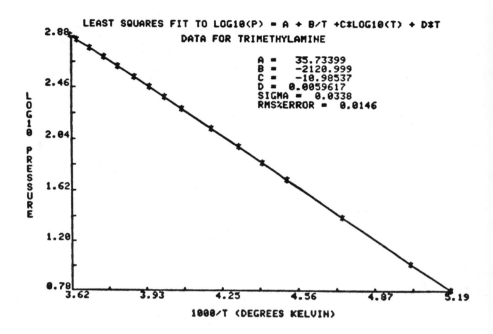

FIGURE 2

LEAST SQUARES FIT TO LOG10(P) = A + B/T +C*LOG10(T) + D*T
DATA FOR TRIMETHYLAMINE

A =   35.73399
B =   -2120.999
C =   -10.98537
D =   0.0059617
SIGMA =   0.0338
RMS%ERROR =   0.0146

LOG10 PRESSURE

1000/T (DEGREES KELVIN)

The problem of numerical integration or quadrature occurs frequently in physical chemistry in which
quite often the function cannot be integrated analytically as in the case of the error function which
appears in kinetic theory in the determination of the
fraction of molecules in a collection having energies
in excess of a specified amount, E^*. Equation (2)
below is derived in Castellan:[4]

$$N(E^*)/N = 2(E^*/\Pi kT)^{1/2}e^{-E^*/kT} + erfc((E^*/kT)^{1/2}). \quad (2)$$

While the above fraction can be readily evaluated for
any single value of the specified energy using tables
of the co-error function, it is quite often of interest to see graphically the conditions under which the
integral term can be ignored over a large range of
specified energies.

FIGURE 3

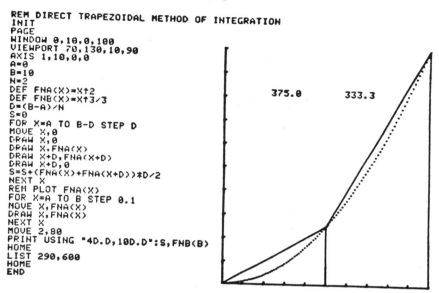

```
REM DIRECT TRAPEZOIDAL METHOD OF INTEGRATION
INIT
PAGE
WINDOW 0,10,0,100
VIEWPORT 70,130,10,90
AXIS 1,10,0,0
A=0
B=10
N=2
DEF FNA(X)=X↑2
DEF FNB(X)=X↑3/3
D=(B-A)/N
S=0
FOR X=A TO B-D STEP D
MOVE X,0
DRAW X,0
DRAW X,FNA(X)
DRAW X+D,FNA(X+D)
DRAW X+D,0
S=S+(FNA(X)+FNA(X+D))*D/2
NEXT X
REM PLOT FNA(X)
FOR X=A TO B STEP 0.1
MOVE X,FNA(X)
DRAW X,FNA(X)
NEXT X
MOVE 2,80
PRINT USING "4D.D,10D.D":S,FNB(B)
HOME
LIST 290,600
HOME
END
```

375.0   333.3

Figure 3 shows the simple graphical trapezoidal
integration program which can be used to integrate any
function of a single independent variable as a function of the number of panels. For most chemical purposes, this simple form of quadrature is sufficient to
provide the required accuracy and has the special pedagogical value of verifying the fundamental theorem of
the integral calculus in the simplest way. The numerical and analytical values of the integral are printed
in the figure from left to right, respectively. The
results of a somewhat more sophisticated version of
this program are shown in Figure 4 in which the left
member of equation (2) is plotted as a function of the
specified energy, E^*. The solid curve ordinates are
the sum of the two terms in equation (2) while the
dotted curve represents the contribution of only the
first term in the right hand member. Clearly, in most
cases the integral term cannot be ignored.

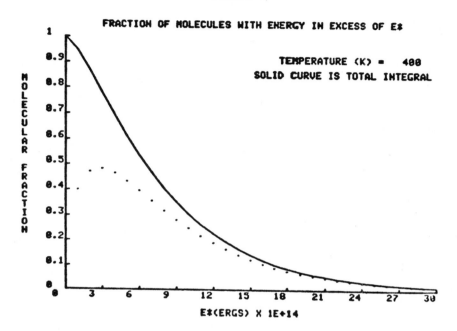

FIGURE 4

FRACTION OF MOLECULES WITH ENERGY IN EXCESS OF E≇

TEMPERATURE (K) =   400
SOLID CURVE IS TOTAL INTEGRAL

E≇(ERGS) X 1E+14

The formulation of many problems in physical chemistry and quantum mechanics in mathematical form leads to one or more ordinary or partial differential equations. In most problems of real interest it is found that the derived equations do not possess a closed form or analytic solution. In chemical kinetics one way around this problem, in part, is the steady-state treatment which is also valuable for the chemical insights it provides for understanding the mechanism of a given reaction. In this treatment, the concentrations of the unstable intermediates are assumed to remain constant during the course of the reaction thereby greatly simplifying the differential equation(s) describing the overall processes. Fortunately, modern digital computers and numerical methods of analysis make it possible to approximate the solutions to such equations to almost any degree of accuracy required. Depending on the given conditions, ordinary differential equations can be classified as either initial value type problems or boundary value type problems. In the former case, the values of the dependent variables and their derivatives are known at a specific point, usually the starting point. In the latter type of problem, the conditions are specified at two or more points. As might be expected, the boundary value problem is more difficult and beyond the scope of the present discussion. Also, any ordinary differential equation of order n can be written as a set of n simultaneous first order equations each of which can be written in the form:[5]

$$dy/dx = f(x,y) = f(x,g(x)) = y'(x) \qquad (3)$$

where y may also be a function of x. Starting with the initial conditions of knowing the slope or derivative at the starting point, x_0, y_0, the next point can be calculated by integrating equation (3):

$$y_1 = y_0 + \int_{x_0}^{x_1} f(x,y)\,dx \simeq y_0 + y'(x)\Delta x \qquad (4)$$

where a convenient approximation for the function under the integral sign is used. In the simple Euler method, the function f(x,y) is assumed to be constant over the interval dx. Geometrically, this amounts to using the point-slope formula of a straight line to predict the next value of y_1 at x_1. Repetition of the process from point to point results in an approximation to the desired integral. The results of applying the simple Euler method to the solution of a differential equation whose analytical solution is also plotted as the solid curve is shown in Figure 5 with the percentage error shown in the right most column below the program for a few points.

FIGURE   5

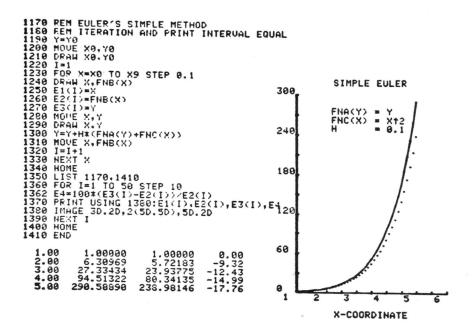

```
1170 REM EULER'S SIMPLE METHOD
1180 REM ITERATION AND PRINT INTERVAL EQUAL
1190 Y=Y0
1200 MOVE X0,Y0
1210 DRAW X0,Y0
1220 I=1
1230 FOR X=X0 TO X9 STEP 0.1
1240 DRAW X,FNB(X)
1250 E1(I)=X
1260 E2(I)=FNB(X)
1270 E3(I)=Y
1280 MOVE X,Y
1290 DRAW X,Y
1300 Y=Y+H*(FNA(Y)+FNC(X))
1310 MOVE X,FNB(X)
1320 I=I+1
1330 NEXT X
1340 HOME
1350 LIST 1170,1410
1360 FOR I=1 TO 50 STEP 10
1362 E4=100*(E3(I)-E2(I))/E2(I)
1370 PRINT USING 1380:E1(I),E2(I),E3(I),E4
1380 IMAGE 3D.2D,2(5D.5D),5D.2D
1390 NEXT I
1400 HOME
1410 END
```

SIMPLE EULER

FNA(Y) = Y
FNC(X) = X↑2
H   = 0.1

1.00	1.00000	1.00000	0.00
2.00	6.30969	5.72183	-9.32
3.00	27.33434	23.93775	-12.43
4.00	94.51322	80.34135	-14.99
5.00	290.58890	238.98146	-17.76

X-COORDINATE

Perhaps the most widely used numerical method of solving differential equations is the fourth order Runge-Kutta procedure, in which the interval x_0, y_0 to x_1, y_1 is divided into four sub intervals and the integral of the function is then calculated as the sum of the integrals over the four sub intervals. The function is taken to be constant over each sub interval as in Euler's method, but by a judicious choice of the points at which the function is evaluated a much higher order of truncation error can be obtained as demonstrated for the same function as above in Figure 6. In this case, it is apparent that the numerical approximation exceeds the graphic resolution of the display where the right most column of data provides the numerical percentage error from the analytical result. In all cases the special advantage of the high resolution interactive graphics is that the user can readily ascertain the goodness of fit and the existence of any discontinuities over the data space.

FIGURE 6

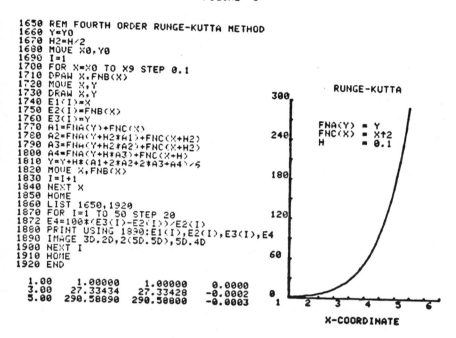

```
1650 REM FOURTH ORDER RUNGE-KUTTA METHOD
1660 Y=YO
1670 H2=H/2
1680 MOVE X0,Y0
1690 I=1
1700 FOR X=X0 TO X9 STEP 0.1
1710 DRAW X,FNB(X)
1720 MOVE X,Y
1730 DRAW X,Y
1740 E1(I)=X
1750 E2(I)=FNB(X)
1760 E3(I)=Y
1770 A1=FNA(Y)+FNC(X)
1780 A2=FNA(Y+H2*A1)+FNC(X+H2)
1790 A3=FNA(Y+H2*A2)+FNC(X+H2)
1800 A4=FNA(Y+H*A3)+FNC(X+H)
1810 Y=Y+H*(A1+2*A2+2*A3+A4)/6
1820 MOVE X,FNB(X)
1830 I=I+1
1840 NEXT X
1850 HOME
1860 LIST 1650,1920
1870 FOR I=1 TO 50 STEP 20
1872 E4=100*(E3(I)-E2(I))/E2(I)
1880 PRINT USING 1890:E1(I),E2(I),E3(I),E4
1890 IMAGE 3D.2D,2(5D.5D),5D.4D
1900 NEXT I
1910 HOME
1920 END
```

RUNGE-KUTTA

FNA(Y) = Y
FNC(X) = X↑2
H   = 0.1

1.00	1.00000	1.00000	0.0000
3.00	27.33434	27.33428	-0.0002
5.00	290.58890	290.58800	-0.0003

X-COORDINATE

In those problems for which similar analytic solutions cannot be examined as above, a variety of error estimating formulas are available for each numerical method of interest.

Although the simple Euler method is seldom used for most practical applications in engineering and physics, its beautiful simplicity makes it attractive pedagogically. Furthermore, the 14 digit numerical accuracy of the 4051 makes it possible to use sufficiently small step sizes to handle a variety of chemical problems within the limits of experimental error.

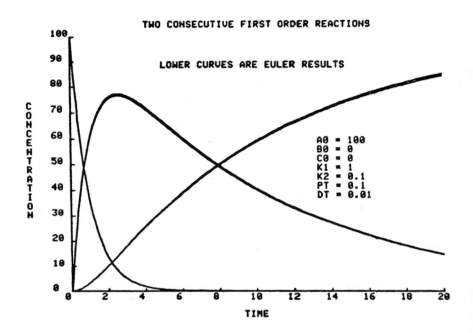

FIGURE   7

TWO CONSECUTIVE FIRST ORDER REACTIONS

LOWER CURVES ARE EULER RESULTS

A0 = 100
B0 = 0
C0 = 0
K1 = 1
K2 = 0.1
PT = 0.1
DT = 0.01

Figure 7 demonstrates rather convincingly that the
simple Euler provides a satisfactory solution to the
simultaneous set of equations describing the progress
of two consecutive first order chemical reactions provided dt is sufficiently small, i.e., 0.01 or less
and the rate constants K1 and K2 are not too large.
Extension of this technique to any set of simultaneous first order equations only requires that their
Euler representations be included within the same in-
crementing loop, and that the initial conditions be
specified.   Likewise, higher ordered equations can be
similarly solved by reduction via appropriate substi-
tutions to a set of first order equations.   Once the
user discovers the power of this approach to solving
differential equations, his progress to more sophisticated methods is assured along with a better understanding of classical analytic methods.

CONCLUSIONS

The commercial availability of moderately priced high resolution graphic micro-computers such as the Tektronix 4051 has added a new dimension to the teaching of numerical methods in physical chemistry. The interactive easy to use graphics provides an exciting visual medium for implementing the self-paced auto-tutorial type of learning of these and other mathematical concepts in a real environment. Concomitantly, the development of a broad range of sophisticated and easy to use general purpose numerical programs by the manufacturer, with continuing library support, encourages the routine application of these powerful methods of data reduction by chemists and other scientific users with a minimum of effort.

REFERENCES

1. V. Fried, H.F. Hameka, and U. Blukis, "Physical Chemistry", MacMillain Publishing Co., Inc., New York (1977) p. 165

2. Pollnow, G.F., J. Chem. Ed. 48, 518 (1971)

3. Aston, J.G., et al, J. Am. Chem. Soc. 66, 1174 (1944)

4. G. W. Castellan, "Physical Chemistry", Second Edition, Addison-Wesley Publishing Co., Reading, Mass. (1971)

5. D. D. McCracken and W.S. Dorn, "Numerical Methods and FORTRAN Programming", First Edition, John Wiley & Sons, Inc., New York (]964) pp. 311-364.

ACKNOWLEDGEMENT

This work was supported by the UW-Oshkosh Faculty Development Program and by the National Science Foundation Grant No. SER 76-16644.

R. C. MORRISON, D. LUNNEY,
M. M. CETERA, R. V. HARTNESS,
G. LOCKLAIR, R. B. RANSOM,
and D. SOWELL   CHAPTER 13

A Calculator Program for a Talking Microcomputer
Designed to Aid Blind Students
in Chemistry Laboratories

I. The System and Its Origins

Introduction. The senior authors of this
paper were not aware of the difficulties of training visually handicapped chemistry students until
they were actually presented with the problem.
When we looked into the commerical availability
of aids for blind chemistry students, we found a
almost nothing. Tactile thermometers, balances,
etc., (1) suitable for use at the elementary and
secondary levels are available, but are not
sufficiently accurate for college work. Even the
most obvious learning aids, for example, three-
dimensional molecular models with tactile identification of atoms, tactile graphs of chemically
important functions, etc., are not commerically
available. Efforts now underway to recruit the
handicapped into the sciences and to train the
handicapped in the sciences at the elementary and
secondary levels (2) will increase the number of
handicapped students who wish to pursue careers
in the sciences, and will increase the need for
aids which give them reasonable access to both
lecture and laboratory experiences. We have
chosen to concentrate our efforts on providing
visually handicapped chemistry students with
laboratory experiences which parallel those of
their sighted peers; this will be of benefit not
just to chemistry majors, but to visually handicapped students in all the natural sciences and
all branches of engineering, because all these
fields require chemistry courses as cognates.
Our approach has been to emphasize instrumental measurements rather than classical wet techniques, for a number of reasons. Many of the

classical experiments done in freshman and sopho-
more laboratories are never done except in these
laboratories (the qual scheme is a good example);
the experimental methods most likely to be encountered again by the student in later work are the
instrumental methods. There is also a clear trend
in the natural sciences toward instrumental measurements and away from direct visual observations;
therefore, anything which helps the handicapped
student use instruments is a step in the right
direction.

Our consideration of the problem led us to
conclude that the most fruitful approach to giving
blind students access to instruments would be the
development of a portable talking laboratory microcomputer with powerful analog and digital input-
output capabilities which would interface readily
to a variety of scientific instruments and would
function as a talking tutorial data acquisition
system. We are now developing just such a system
and writing software for suitable undergraduate
chemistry experiments. The purpose of this paper
is to describe the system we envision and to report
our progress to date.

The Ultimate Personal Computer. Because the
system is intended to interface easily to as many
instruments and sensors as possible, flexible input-
output capabilities are essential. Analog inputs
will include a high impedance channel, a high gain
differential input, an isolated input, and current
inputs. Analog outputs will include both current
and voltage.

Digital inputs and outputs will include the
usual serial RS-232 for communication between the
system and other computers or for driving an
external braille printer, and the IEEE 488 data
bus (3). (The IEEE bus is now rare on chemical
instruments, but is common on electronic instruments.) Other input-output functions will include
relays for switching 120-volt AC loads, a micro-
cartridge tape drive, and floppy disc interface for
use with an external flexible disc drive. The
principal communication channels between the user
and the machine will be keyboard input and voice
output.

In addition to its obvious central function
as a collector and analyzer of data and as a talking laboratory tutor, the system will be able to
function as a talking scientific calculator, as a
talking computer terminal, and as a stand-alone
computer. The power and flexibility of this system will make it useful not only in an instructional setting but also in research and industrial
laboratories, indeed, wherever observations are
made through the medium of instruments. Its flexibility and wide general utility induced us to call
it a Universal Laboratory Training and Research
Aid (ULTRA).

Realizing the ULTRA System. We have chosen
to realize the ULTRA system using the Z-80 board
family manufactured by Zilog Corporation. This
family of subsystems is available on compact 7.5 by
7.75 inch cards, has modest power dissipation (about
10 watts per board), includes a 16-channel data
acquisition board, and uses the powerful Z-80 processor. Furthermore, available software includes
FORTRAN, maximizing the portability of the software,
as will be discussed later. We avoided the use of
hobby computers for reasons of reliability; the
visually handicapped scientist or student who uses
this system will not be nearly so tolerant of downtime as a hobbyist would. Also, use of an indus-
trial-grade computer system probably assures better
hardware and software support from the manufacturer.
To facilitate development of experiments, software, and data acquisition circuits, we have acquired a Zilog MCZ-1/20 computer system with dual
flexible disc drives and disc operating system
software. The portable laboratory system will
contain a board set which will be a subset of the
boards contained in the development system; hence,
if software runs on the development system it must
also run on the prototype portable system.
Prototypes of much of the analog signal-
treatment circuitry have been built; the input-
output facilities now available on the development
system are shown in Table I.
The system's analog inputs fall into two broad
groups: Those which accept signals from trans-
ducers or sensors directly (for example, the high

impedance pH input), and those which interface to external instruments. Inputs intended to be used with an instrument include an isolation amplifier for use with systems in which high common mode voltages are present or in which ground loops might cause excessive 60-Hz noise. A voltage-controlled

TABLE I.   Input-Output Capabilities of the
            Ultra Development System

  I.   Man-Machine Interfaces

       A.   Keyboard Input
       B.   Voice Output (Phoneme-Based)
       C.   Voltage-Controlled Oscillator
            (Audible Null Detector)

 II.   Analog Input (16 Channels)

       A.   Isolation Amplifier
            (Computer-Controlled Gain)
       B.   High Impedance (pH, etc.)
       C.   Current (Computer-Controlled Ranging)
       D.   Spares (-10 to +10V)

III.   Analog Outputs

       A.   Voltage (+10V, $\pm$100 mA MAX)

 IV.   Digital I/O

       A.   Serial RS-232-C
       B.   Events Counters
       C.   Spare Parallel Ports

---

oscillator is included to function as an audible null detector; such an output is very useful when an instrument must be set to a given level, as in setting 0 and 100% transmittance on a spectrophotometer. The system provides voice echo of characters entered on the keyboard to allow for error correction.

Speech Output. Speech output is based on a Votrax® model VSK voice synthesizer. This synthesizer generates speech from sound elements called phonemes; with its repertoire of 61 phonemes it can utter any word in the English language. This mode of speech synthesis was chosen because it is very economical of computer storage; phoneme-based speech output requires only about a hundred bits per second of speech. Each phoneme is represented by a six-bit code, so the synthesizer can be driven by an eight-bit output port. The synthesizer can be silenced by applying a null code to its input.

The VSK synthesizer has an internal clock which times speech synthesis and provides handshaking with the input-output port. When the synthesizer is not generating phonemes (i.e. when it has been silenced by applying the null code) the clock runs free with a period of about 40 ms. The clock output consists of a train of negative-going pulses at CMOS levels; the pulses are about 0.5 ms wide. When the synthesizer begins generating a phoneme the clock goes high and remains high until the phoneme is completed, whereupon a negative-going pulse is produced. In our system the pulse generates an interrupt which causes the code for the next phoneme to be applied to the synthesizer about 30 μs after the rising edge of the pulse. Use of an interrupt driver allows the microcomputer to perform other tasks while the synthesizer generates phonemes, whose duration is typically a few hundredths of a second. Since the Z-80 runs at a clock frequency of about 2.4 MHz the processor can execute many instructions between phonemes.

Not all of the 64 possible phoneme codes represent distinct sounds. Two of the codes represent pauses, and one is a null. The remaining 61 codes represent vowel and consonant sounds; the vowel sounds may have different time lengths. Each phoneme is represented by a short mnemonic symbol; the time length of a vowel is represented by a numeric suffix. For example, the phonemes UH, UH1, UH2, and UH3 all refer to the same phonetic sound; the UH phoneme has the longest duration and UH3 the shortest.

Spoken words are produced by sequentially sending the appropriate binary codes to the synthesizer.
For example, the synthesizer would be made to say
"absorbance" by sending it the phonetic codes

AE1	EH3	B	S	O1	R	B	EH3	N	S
2F	00	0E	1F	35	2B	0E	00	0D	1F

The numbers below the phoneme symbols are the hexidecimal representations of the binary codes that
would be sent to the voice synthesizer.
      The strings of phonetic codes for the words
and scripts used with our software are developed
by using a program which allows us to create, hear,
and edit disc files containing the phonetic codes
for the scripts. We listen to and edit scripts
until they are reasonably intelligible. The program then creates the FORTRAN source code that will
cause the script to be said. The FORTRAN code is
stored on a disc file that can be retrieved later
by the disc file editor for inclusion in software
that is being developed. An interesting aspect of
writing phoneme codes for voice synthesis is that
the spoken output has the accent of the programmer.
For this reason, programmers with severe regional
accents do not produce acceptable results.

II.   An Important Software Example

A Talking Calculator Program. In this section
we will describe a calculator program with keyboard
input and spoken audio output which has been developed for interactive use with the ULTRA system.
The calculator program can be linked with other
programs intended to aid blind students with experiments in undergraduate chemistry laboratory
courses, or can be invoked simply for use as a
calculator.
      We have written as much of our software as
possible in FORTRAN so that we would not be tied
to any one microcomputer. The calculator program
is written in FORTRAN and is currently running on
an HP2100 minicomputer without voice output and on
a Zilog MCZ-1/20 microcomputer with voice output.

The only assembler routine used by the program is
the input/output driver used for voice output
through the voice synthesizer. Hence, the program
is transportable to other microcomputers having
FORTRAN IV. The FORTRAN language appears to be
adequately efficient; we obtain less than one
second response time for most calculator statements.
    The arithmetic operators and special functions
listed in Table II are those which we thought
would be useful in undergraduate chemistry laboratories and which would normally be found on most

Table II. Calculator Functions

Symbol	Definition
+	Addition
-	Subtraction
*	Multiplication
/	Division
∧	Exponentiation
=	Variable Storage
:	Array Creation
( )	Expression Grouping
[ ]	Subscripting
#	Array or Variable Deletion
LOG	Common Logrithm
LN	Natural Logrithm
EXP	Exponential Function
COS	Cosine
SIN	Sine
SQR	Square Root
PLOT	Least Squares Calculation
SAY	Say Values of an Array

scientific calculators. The calculator has 50
memory locations that may be used by the student for
storing constants or intermediate results. Experimental data may also be placed in these locations by
other programs to which the calculator is linked.

An arithmetic expression is entered as an algebraic expression and may contain arithmetic operators, constants, variables or arrays. The result is said by the computer unless it is stored using the "=" or ":" operators. If an array appears in an arithmetic expression, the result should be stored in an array using the "=" operator or the ":" operator.

Variable names and array names are used to access the data storage area. Each variable or array name consists of one to fifteen alphabetic characters. They may be used in arithmetic expressions wherever numeric values would normally be used. If two or more arrays appear in an arithmetic expression, they must have the same length.

Elements in an array may be used in a calculation by using brackets to subscript the array. For example, the second element of an array called Z would be accessed by Z [2]. A subscript may be a number, variable, subscripted variable, or an arithmetic expression. If the subscript is a number the brackets may be omitted.

The operator hierarchy is the same as it is in FORTRAN, where operations are performed in the following order:
1) function evaluation,
2) exponentiation,
3) division and multiplication,
4) addition and subtraction.
Evaluation of an arithmetic expression proceeds from left to right within a hierarchical level.

The calculator has two types of statements, examples of which are given in Table III. They are the data storage statements and arithmetic statements.

The data storage statements which use the "=" or ":" operators are used to store constants in up to fifty memory locations. The "=" may be used to create arrays from arithmetic expressions containing existing arrays. For example, if T is an array containing six elements, the expression A=-LOG(T) would create an array, A, containing six elements which correspond to the negative of the common logarithms of the elements of the array T. The ":" operator is used to create arrays where several numbers, variables, arrays, or values of arithmetic

Table III. Sample Calculator Statements

Examples of Arithmetic Expressions

Expression	Result
1+2*3+2∧2	11
-LOG(10∧3)	-3
LINE=PLOT(T,A)	Use a least squares procedure to find the slope and intercept of a straight line plot or T vs. A. The slope is stored in LINE 1 and the intercept is in LINE 2.

Examples of Data Storage Statements

Statement	Result
P=3.1416	3.1416 is stored in P
Z:1.0, 0.5, P, LOG(2)	Creates Z with elements 1.0, 0.5, 3.1416, and .301

---

expressions are entered into the array. The numbers, variables, etc. are separated by commas in the data storage statement.

The arithmetic statements do not contain the "=" or the ":" operators, but consist of simple arithmetic expressions whose values are computed and given as spoken output.

The calculator allows the student to process experimental data interactively as it is being collected. For example, a student could collect four or five data points to obtain a Beer-Lambert plot of absorbance vs. concentration. After collecting the data the calculator can be used to invoke a least squares routine to find the slope, intercept and standard deviation of the best straight line through these points. The student can decide to use these results if they are satisfactory, or can repeat some of the measurements or make measurements on new samples. This gives the blind student the same flexibility that sighted students have in being able to "look" at the data and do the calculations while still in the labora- tory, and offers the same opportunity to make mis- takes if incorrect expressions are used with the calculator.

Implementation Techniques. The main program calls a routine to get a line from the keyboard, a routine to parse the line, a syntax checking routine, and the CLCTR routine which controls flow through the calculator subroutines. The parsing and syntax checking routines are used for command lines for other programs used to run experiments as well as for the calculator lines. An overview of the program flow is given in Figure I.

The calculator statements are entered in free format. The parsing routine sets up operator and operand codes and pointers for each item in the calculator source line. Constants are evaluated and stored in a working storage area.

The early operator reverse Polish notation(5) is used to facilitate expression evaluation. We have used a standard algorithm(5) for translating the arithmetic expression in the source line to the

Figure I-A.   Flow Chart of the Main Calculator
Routine

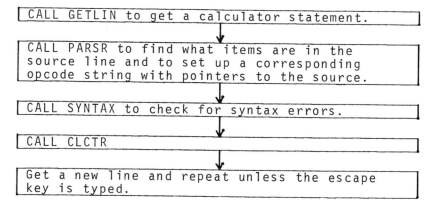

| CALL GETLIN to get a calculator statement. |

| CALL PARSR to find what items are in the source line and to set up a corresponding opcode string with pointers to the source. |

| CALL SYNTAX to check for syntax errors. |

| CALL CLCTR |

| Get a new line and repeat unless the escape key is typed. |

Figure I-B.   Flow Chart of CLCTR, the Calculator
              Driving Routine.

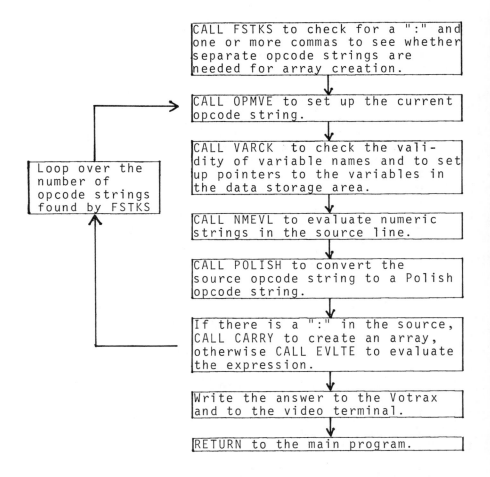

Polish string; however, operators must first pass
to an operator stack. Each operator's order of
placement on the Polish string depends on its
hierarchical level as well as its position in the
source line.
    The expression is evaluated by scanning the
Polish string from left to right. Values of

operands are placed on a value stack. A dyadic
operator encountered in the Polish string operates
on two values in the value stack with the result
being placed back in the value stack. An "="
operator or a ":" operator in the Polish string will
cause the result of the expression to be stored in
an array or variable, otherwise the result is given
as spoken output.

   <u>Linkage to Other Programs</u>. The calculator
program may run as an independent program or it
may be linked to other programs to provide interactive calculations on experimental data. The
program may be run independently by simply providing
a main program that calls the input, parsing,
syntax checking and calculator driving routines.
   We have linked the calculator program to a
program that is used to aid blind students in
obtaining data from a spectrophotometer such as
a Bausch and Lomb Spectronic-20. This spectrophotometric program may be used to search for a wavelength giving maximum absorbance, verify the Beer-
Lambert Law, or do a colorimetric analysis of an
unknown. The calculator program is used to aid in
carrying out the colorimetric analysis of an
unknown sample. The concentrations of standards
are entered through the keyboard and are stored in
the calculator storage area. The student is responsible for calculating absorbances using the equation,
A=-LOG(T), and for knowing that a graph of A vs. C
should be a straight line from which unknown concentrations can be determined.
   The student invokes the calculator routines by
entering the appropriate command on the keyboard.
For example, the values of the transmittances and
concentrations could be stored in the calculator
storage area in arrays labeled T and C respectively.
The student would enter the calculator statement,
A=-LOG(T) to create an array of absorbances. He
would then enter the statement LINE=PLOT (A, C) to
obtain the slope and intercept of the least squares
fit of a straight line to the experimental data.
The slope and intercept are stored in variables
LINE 1 and LINE 2 respectively, and are available
for subsequent determinations of unknown concentra-

tions. If additional standards are run, the process may be repeated.

Acknowledgement. This work was supported by a grant from the Bureau of Education for the Handicapped.

References

1. American Foundation for the Blind, International Guide to Aids and Appliances for Blind and Visually Impaired Persons, 1977.
2. E. Walsh, Science, 196, 1424 (1977).
3. Institute of Electrical and Electronics Engineers, IEEE Standard Digital Interface for Programmable Instrumentation, (Standard 488-1978), 1978.
4. D. E. Tallman, J. Chem. Educ., 55, 605 (1978).
5. Harry Katzan, Jr. Advanced Programming, New York, Van Nostrand Reinhold Company, 1970, Chaper 4.

Microprocessor Delivery of PLATO
Instructional Material

The On-Line PLATO System. The University of Illinois
PLATO system (1), which now has 1000 terminals connected to a
dual main frame CDC CYBER computer system which provides a
million terminal hours per year, has been used for the past six
years to investigate the use of computer-assisted instruction.
A classroom located in the Chemistry Department containing 32
PLATO terminals connected to the central computer system by
microwave has been used by up to a thousand students/semester
doing required work in courses ranging from general chemistry
to graduate level work in organic chemistry.
    The computer is used to provide direct instruction (2) in a
general chemistry course where students work about 2 hours per
week at a terminal in addition to 1 hour of lecture, one quiz
section and a laboratory session. Organic chemistry students
work 2 to 3 hours per week on PLATO in addition to 3 lectures
per week. It has been found that highly interactive lessons
which integrate simulated experiments, animations, dialog and
practice problems receive strong student acceptance. For
example, 92% of the students in one sample said that they would
recommend taking the course with PLATO as opposed to the course
without CAI. In addition, interactive programs have been shown
to result in better student performance in the laboratory (3).
    The on-line system provides easy access to a very large
library of courseware, automatic terminal and classroom management, computer managed instruction, shared data bases for on-
line grade books, student response collection and analysis for
optimizing new instructional material and inter-terminal
communication features.
    However, the on-line system requires a communication system
with the equivalent of 1200 baud per terminal and access to a
large main frame PLATO computer system. Since many users of

177

instructional materials are remote from a central system or do
not have extensive need for the unique on-line features it
seemed desirable to develop an off-line system which has the
capability of delivering the same instructional material which
was found to be educationally effective with the on-line system.

The Off-Line PLATO System. The display requirements for an
off-line system used for providing computer-assisted instruction
in chemistry include both upper and lower case and superscripts
and subscripts. This makes it easy to display and have the
student enter chemical formulas such as $NaNO_3$. High resolution
vector graphics make it possible to draw structural formulas,
plot functions, and show IR and NMR spectra. High resolution
graphics also makes possible detailed animations which have been
found to help students visualize the course of chemical reactions. In addition, the ability to design and display your own
character set makes it possible to use characters and symbols
which are unique to a particular subject area. The use of touch
sensitive displays also facilitates easy student interaction
with the programs.

In addition, an off-line system must provide sufficient
computer capacity to judge input from students in the form of
either numbers or complete sentences within a few tenths of a
second and to respond to the students' work with text, figures
and diagrams.

Initial experiments in delivering PLATO instructional
material off-line have involved a PLATO V terminal (4) which is
equipped with an 8080 microprocessor and 16K bytes of read-only
memory (ROM) and 8K bytes of random access memory (RAM) and a
single density, single sided, 8 inch floppy disk (512K bytes).
The ROM contains a μ-TUTOR interpreter.

Instructional material is written on-line to CYBER at the
present time, although off-line editing is expected to be made
available later. Programs are written in μ-TUTOR with the
standard TUTOR editor the same way that regular PLATO lessons
are normally prepared. The PLATO system then generates compiled
code which is "down loaded" to the terminal and stored on an
attached diskette for later use by students. The use of binary
code stored on the disk results in rapid execution times, and
the increased band width from 120 bytes/sec when connected to a
central computer system to 30,000 bytes/sec between the disk
and the terminal greatly reduces the time to generate displays.
Text is plotted at 3,000 characters per second and lines may be
drawn at rates of 100 to 1,000 per second, depending on their
length. Execution times for 14 significant-figure floating-
point arithmetic are: add, 1 ms; multiply, 4 ms; and divide,
9 ms. The terminal automatically pages the information from the

disk which is needed at any given time to run the program. This makes it possible to use instructional material whose length is determined by pedagogical considerations instead of the size of the terminal memory. Typical delays in paging information from the disk are less than 0.1 sec.

Some features of μ-TUTOR (5) are illustrated in this sample program.

```
μtutor
define   i,8:chars(30)
i,16     i,16:long,right,wrong,nprobs
         f,48:P,V,T,n
         R=0.08205
charset  gases,chars
*
unit     press
next     prob
at       1010
write    If you were to push down on this piston what
         would happen to the pressure of the gas inside?
at       1528
write
```

```
draw     216,272;216,210;240,210;240,274
arrow    2226;chars(1),30
wrong    (down,dn,lower,decrease)
.        at      2510
.        write   No, if you compress the gas the pressure
.                increases.
answer   (up,increase,increases,higher)
.        at      2510
.        write   Right!  Compressing the gas increases the
.                pressure.
no
.        at      2510
.        write   The pressure either increases or decreases.
endarrow
*
unit     prob
back     press
randu    P,5
randu    n,2
randu    T,100
calc     V←nR(T+273.2)/P
```

```
at       1010
write    What is the volume of ⟨s,n⟩ moles of an ideal gas
         at ⟨s,T⟩°C and ⟨s,P⟩ atm
arrow    1425;chars(1),10
ansv     V,10%
.        at      1810
.        write   That is right!
no
.        at      1810
.        write   Calculate V by substitution of the
.                values given into the formula PV = nRT
endarrow
```

The -define- statement allows the author to set up 8 and 16 bit integer and 48 bit floating point variables and arrays with convenient names. Progression from one "unit" to the next is controlled by the -next- and -back- commands which also have conditional forms. Text is written on the screen with the -write- command at the position indicated by -at- which in these examples indicates the line and character position on the screen. The piston is created with user defined characters while the cylinder is drawn with a -draw- command which draws lines between the indicated coordinate positions on the screen. Correct student responses are indicated with the -answer- command for verbal answers and the -ansv- for numerical responses. The -wrong- allows the specification of anticipated wrong answers which, if given, result in the execution of the immediately following indented code. At present, the μ-TUTOR language has over 60 commands, so those illustrated here represent just a small subset of those available. Because of the structure of μ-TUTOR, a very large numer of commands are possible.

A few displays which illustrate some of the things that can be done off-line are illustrated in Figures 1-3. The program (6) illustrated in Figure 1 deals with spin-spin coupling in proton magnetic resonance spectroscopy. In the display illustrated the student is given the option of exploring the relationship between J and δ by suggesting values for an A_2B system and having the μ-processor calculate and display the corresponding spectrum.

In Figure 2, a student has suggested a synthesis of n-propyl alcohol from ethanol by touching the reagents shown on the screen. The program (7) works on a general functional group analysis so that alternative procedures will be automatically accepted.

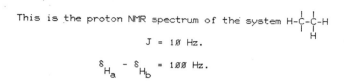

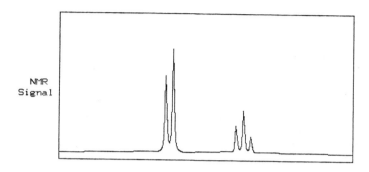

Magnetic Field

To plot another spectrum press NEXT.

Figure 1.   One display from a lesson on spin-spin coupling
in proton NMR. Here the student specifies the
coupling constant and the chemical shift for an
A_2B system and the μ-processor calculates and
displays the spectrum.

The collection of data for the experimental determination
of the ideal gas laws (8) is illustrated in Figure 3. Here the
student collects pressure and volume data at constant temperature from which plots of P vs. V will be generated.

To facilitate easy use of the system by students, a program
may be written which routes students to the appropriate lessons
on the disk. One way this feature may be used is to generate
an index of lessons which is displayed when the system is turned
on or, if desired, when the student carries out an appropriate
sign-on procedure. Lessons may be accessed in about a second
from such an index by a single key press. Of course, at any
point in a lesson the student may elect to return to the router
index to select another activity.

Synthesis of: $CH_3CH_2CH_2OH$

CH_3CH_2OH $\xrightarrow{PBr_3}$ CH_3CH_2Br $\xrightarrow{Mg\ Et_2O}$

CH_3CH_2MgX $\xrightarrow{HCHO}$ $CH_3CH_2CH_2OH$

Touch a Reagent

PBr_3	$SOCl_2$	HBr	H_2SO_4	H_3O^+	CrO_3	$KMnO_4$
NaOH H_2O	KOH EtOH	a LAH b H_3O^+	a B_2H_6 b H_2O_2	Mg· Ether	a CO_2 b H_3O^+	a EtMgBr b H_3O^+
a CH_3MgBr b H_3O^+	HCHO	CH_3CHO		Start over	HELP	DONE

Figure 2.   In a lesson on multi-step synthesis students suggest ways to prepare a particular compound by touching the reagents displayed on the screen. The program uses a functional group analysis so that it is not necessary to program all possible ways to solve a given problem.

Data sets may also be set up on the disk to record data generated as the student works. Such data sets may also be used to store experimental data entered through the external jacks of the terminal from, for example, experiments which are controlled by the microprocessor.

This work suggest that it is possible to deliver highly interactive, educationally effective instructional lessons off-line with a microprocessor based terminal which has high resolution (512 x 512) vector graphics.

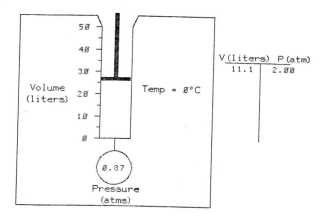

Figure 3. In a lesson on ideal gases students collect data in a simulated experiment. The student's data are then used to develop the ideal gas laws.

References

(1) S. G. Smith and B. A. Sherwood, Science, 192, 344 (1976)
(2) R. Chabay and S. G. Smith, J. Chem. Ed., 54, 745 (1977)
(3) C. Moore, S. G. Smith and R. A. Avner, J. Chem. Ed., 57, 196 (1980)
(4) J. Stifle, S. G. Smith and D. Andersen, Proceedings Association for the Development of Computer-Based Instructional Systems, 3, 1027 (1979)
(5) μTUTOR is being developed at the University of Illinois by David Andersen and co-workers.
(6) S. G. Smith, "NMR Spin-Spin Coupling", Control Data Corporation, 1972
(7) A PLATO lesson by S. G. Smith, copyright, 1979

(8) S. G. Smith, "Properties of Ideal Gases", Control Data
     Corporation, 1976

CHAPTER 15

Microcomputer Based Computer Assisted Instruction

The principal tenet of this article is that computer assisted instruction, (CAI), can be effectively implemented on a low-cost "consumer" computer. My primary interest is the implementation of various forms of CAI on one or another microcomputer configuration; I do not intend to discuss the pedagogical value of any particular implementation, but rather to illustrate that virtually every known form of computer enhanced education, whether computer assisted instruction, computer managed instruction (CMI), or computer assisted learning (CAL) is accessible on a computer system which is assembled, tested, and packaged in a convenient and usually attractive as well as robust form, and available to the end user for $900 to $2,000. I shall use examples from IIT and from a typical microcomputer system to show that the design considerations behind personal computers have lead to superior delivery systems for computer enhanced education. I wish to direct the interested reader to a continuing series of articles on computers in chemical education, appearing in The Journal of Chemical Education under the general editorial supervision of John Moore (1-6). The superiority of "foolproof" stand-- alone microcomputers extends beyond the end user to the instructor who may have been inhibited from using

computers by some of the nuances of general purpose
machines. In the following exposition, I shall
compare and contrast computer enhanced chemical education using a central "mainframe" computer with the
same tasks undertaken on a typical microcomputer
system. I begin with a brief discussion of our experience with one central computer system, the IIT
PR1ME 400.

THE MAINFRAME COMPUTER IN CHEMICAL EDUCATION.
At IIT, we support 16 hardwired terminals, selected
principally on considerations of cost and (presumed)
durability in a student environment. They were also
part of the system package. These terminals display
24 lines of 80 characters, and are wired to
communicate with the computer at 4800 baud. They do
not support lower case display, the cursor may not be
independently addressed, and the screen can not be
cleared. Programs designed for the IIT terminals are,
therefore, required to halt the scrolling display
every 24 lines to permit the student to read from the
screen. Each topic brought before the student is
written onto the display from the bottom, as the
previous display scrolls off the top, much like an
electronic teletype. Unlike a teleprinter, however,
information scrolled off the top of the screen is lost
to the student, and must be refreshed (scrolling from
the bottom) each time it is needed for reference.
     At IIT, I have installed a large number of in-
teractive tutorial and drill and practice routines for
the General Chemistry program. One of my initial
concerns in setting up a CAI project was a clear
separation of computer science from CAI. Thus, one of
the first routines written was a calling program which
isolates the user from the workings of the computer.
The IIT implementation of the PR1ME 400 operating
system permits us to use LOGIN CHEM as the principal
entry command, TUTOR as a student ID, and US as the
password. The student then types BASIC CHEMISTRY,
which calls the BASIC interpreter and loads and executes the file named CHEMISTRY.
     PR1ME BASIC lacks some of the error trapping
routines which are essential for student use. As an
example, if a student is asked for the percentage of
hydrogen in a hydrocarbon, and inputs 3.6%, the run is
terminated with a bad input error (mixed mode), and

the student is left in the operating system. I have
shown a routine written by E.A.Mottel to trap these
errors. It accepts all input as alphanumeric strings,
then strips off the numerical answer for processing by
the main program. It is an illustration of the
lengths to which we have gone to suppress student
frustration by little things which "computer jocks"
accept without qualm. These barriers frustrate not
only student users, but also competent teachers who
are unwilling or unable to contend with the nuances of
general purpose computers. This subroutine is also
interesting because the problem is not unique to the
PR1ME system, and because the coding is typical of
BASIC programming on many large computers.

```
100 REM ALPHAN SUBROUTINE ALPHA TO NUMERIC CONVERSION
105 REM VERSION 1.0   JUNE 27,1978
110 REM INPUT Z$ OUTPUT O6
120 DIM I$(10)
130 FOR O1=0 TO 9
140 READ I$(O1)
150 NEXT O1
160 DATA '0','1','2','3','4','5','6','7','8','9'
170 PRINT 'INPUT NUMBER'
180 INPUT Z$
190 O2=LEN(Z$)
200 O3=0
210 O4=1
220 O5=0
230 O6=0
240 O7=0
250 O8=0
260 O9=0
270 IF SUB(Z$,1)<>'-' THEN 300
280 O4=-1
290 O5=1
300 O5=O5+1
310 IF O5>O2 THEN 450
320 O$=SUB(Z$,O5)
330 O1=0
340 IF O$=I$(O1) THEN 410
350 O1=O1+1
360 IF O1<10 THEN 340
370 IF O$='E' THEN 450
380 IF O$<>'.' THEN 300
390 O8=1
```

```
400 GOTO 300
410 06=06*10+01
420 09=09+1
430 03=03+08
440 GOTO 300
450 IF 09>0 THEN 470
460 06=1
470 06=04*06
480 04=1
490 05=05+1
500 IF 05>02 THEN 610
510 0$=SUB(Z$,05)
520 01=0
530 IF 0$=I$(01) THEN 590
540 01=01+1
550 IF 01<10 THEN 530
560 IF 0$<>'-' THEN 490
570 04=-1
580 GOTO 490
590 07=07*10+01
600 GOTO 490
610 07=04*07-03
620 05=1
630 IF 07=0 THEN 790
640 IF 07>-39 THEN 670
650 PRINT 'BAD INPUT-UNDERFLOW-NUMBER TOO SMALL'
660 GOTO 690
670 IF 07+09<39 THEN 720
680 PRINT 'BAD INPUT-OVERFLOW-NUMBER TOO LARGE'
690 PRINT 'RE-ENTER YOUR NUMBER':
700 INPUT Z$
710 GOTO 190
720 FOR 01=1 TO ABS(07)
730 05=05*10
740 NEXT 01
750 IF 07<0 THEN 780
760 06=06*05
770 GOTO 790
780 06=06/05
790 PRINT 06
800 GOTO 170
810 RETURN
```

We have experienced little difficulty at the stu-
dent-computer interface. From the foregoing, it
should be clear that I do not want to devote time and

effort to instructing the beginning student in <u>any</u> commands of the type OLD, NEW, RUN, CLEAR, COMPILE, PACK, etc, much less in programming. My purpose is to have the student interact with the program as quickly and easily as possible; after all, my time and effort went into producing an instructional program. These examples illustrate our efforts to isolate computer enhanced chemical education from computer science. It is obvious that these measures introduce considerable overhead into instructional programming.

    <u>The Microcomputer in Chemical Education.</u> I believe that it is possible to overcome these obstacles with a "consumer" microcomputer system. For example, in the simplest configurations, "consumer" computers are ready to run when turned on (thus, in contrast with the IIT system, one need not even call BASIC). Most can be loaded with a one word (or mnemonic) command, and programs executed with the RUN command.

    Typical microcomputers include, but are not limited to, the Radio Shack TRS-80, the Commodore PET 2001, the Apple II, and the Exidy Sorcerer. These computer systems contain the central processor, random access memory (user programmable), an operating system and (usually) the BASIC programming language in read only memory, facilities for reading from or writing to cassette mass storage, and a keyboard in a single "black box". Display is usually provided <u>via</u> a separate video monitor or television receiver. Each of these computer systems is available with a wide variety of extensions, including expanded programmable memory, disk mass storage, hard copy devices, communications hardware, and a number of "trick" extensions for music, speech synthesis, control of other devices, etc. I am particularly interested in the educational applications of the basic product, since it appears, on the surface, to be most cost effective and also the simplest to use. While disk systems offer greater speed in loading, larger storage, and often greater flexibility in the operating system, they also introduce greater complexity in use, and add mechanical components with attendant maintenance problems. Moreover, most of the tutorial programs run efficiently on a microcomputer without disk support. Finally, disk systems add about $600 to the base price

of a microcomputer system, or about 50% of the
purchase price of each work station. Thus, as I see
the cost/effectiveness issue, large scale CAI is most
easily justified when each work station represents the
smallest capital investment consistent with system
performance as it is seen by the user. By that I mean
to say graphics capabilities, adequate error trapping
(as a part of the operating system), adequate keyboard, and easily read display are more important to
the end user than details of the operating system,
file handling strategies and other features which
should be transparent to the student user.

  Microcomputer Assisted Instruction. I have
translated most of the IIT program package from PR1ME
BASIC (a relatively primitive version) into Microsoft
BASIC, which, in its extended forms, is the principal
language of the microcomputer industry (7). Microsoft
BASIC is quite powerful, supporting a complete library
of math functions (matrix manipulation may be optional), powerful logical operations and powerful string
manipulation commands. Minor variations exist within
this product; these variations usually appear in the
I/O routines, and may reflect the market chosen by the
hardware vendor. Thus, the formatted print options
appear to differ between game and graphics oriented
computers on one hand, and business oriented computers, on the other. Here is a portion of a TRS-80
listing. I admit that it does not conform to any
standard of good programming practice, but it does illustrate the stark contrast between "mainframe" BASIC
and "microcomputer" BASIC.

```
210 CLS:PRINT:PRINT:PRINT"IF AN ATOM HAS ";E;" ELECTRO
NS ";:IF E>2 THEN 220ELSEO$="IS":GOTO390
220 IF E>4 THEN 230ELSE O$="2S":T=E-2::GOTO 390
230 IF E>10 THEN 240ELSE O$="2P":T=E-4:GOTO 390
240 IF E>12 THEN 250ELSE O$="3S":T=E-10:GOTO 390
250 IF E>18 THEN 260ELSE O$="3P":T=E-12:GOTO 390
260 IF E>20 THEN 270ELSE O$="4S":T=E-18:GOTO 390
270 IF E>30 THEN 280ELSE O$="3D":T=E-20:GOTO 390
280 IF E>36 THEN 290ELSE O$="4P":T=E-30:GOTO 390
290 IF E>38 THEN 300ELSE O$="5S":T=E-36:GOTO 390
300 IF E>48 THEN 310ELSE O$="4D":T=E-38:GOTO 390
310 IF E>54 THEN 320ELSE O$="5P":T=E-48:GOTO 390
320 IF E>56 THEN 330ELSE O$="6S":T=E-54:GOTO 390
```

```
330 IF E>70 THEN 340ELSE O$="4F":T=E-56:GOTO 390
340 IF E>80 THEN 350ELSE O$="5D":T=E-70:GOTO 390
350 IF E>86 THEN 360ELSE O$="6P":T=E-80:GOTO 390
360 IF E>88 THEN 370ELSE O$="7S":T=E-86:GOTO 390
370 IF E>102 THEN 380ELSE O$="5F":T=E-88:GOTO 390
380 O$="6D":T=E-102
390 PRINT"HOW MANY ELECTRONS":PRINT"ARE IN THE ";O$;"
ORBITAL";:INPUT A
```

In particular, note the extensive use of multiple
statement lines, and IF..THEN..ELSE logical operations. Here's another example, the calculator subroutine I incorporate in programs which require some
calculations.

```
240 PRINT"DO YOU WANT TO USE MY CALCULATOR";
250 INPUTZ$:IFLEFT$(Z$,1)="Y"THEN260ELSEINPUT"PLEASE T
YPE YOUR ANSWER";A:RETURN
260 INPUT"MULTIPLICATION, DIVISION OR SQUARE ROOT (TYP
E M,D,OR S)";Z$:IFZ$="S"THEN270ELSEIFZ$="M"THEN290ELSE
IFZ$="D"THEN300ELSEPRINT"DO YOU REALLY WANT MATH HELP"
:GOTO250
270 INPUT"OF WHAT NUMBER";S:IFS<0THENPRINT"I CAN'T COM
PUTE THE SQUARE ROOT OF A NEGATIVE NUMBER. TRY AGAIN":
GOTO 270
280 PRINT"THE SQUARE ROOT OF ";S;" IS ";SQR(S):GOTO 31
0
290 INPUT"GIVE ME THE FIRST NUMBER";A1:INPUT"NOW THE S
ECOND NUMBER";A2:PRINTA1;" * ";A2;" = ";A1*A2:GOTO310
300 INPUT"GIVE ME THE DIVIDEND";A1:INPUT"NOW GIVE ME T
HE DIVISOR";A2:PRINTA1;" / ";A2;" = ";A1/A2
310 PRINT 512,CHR$(30);"DO YOU NEED MORE HELP";:GOTO025
0
```

I use the tightly programmed BASIC for a couple
of reasons. First, I like it. Second, and more
useful, is that it conserves storage, often at a
premium in a small machine, and it also speeds up execution times. In this tightly packed code, I have
succeeded in running the largest program currently
supported in the IIT Chemistry tutorial package on a
16 Kbyte microcomputer. I have, of course, also
succeeded in compressing most of the other programs in
the package. Nonetheless, I view about 6k bytes as
the useful minimum for most of my programs, not
because I require extra bytes for intricate computing,

but because I like to include "personal" responses to
the student which take up a couple of kilobytes of
string storage. Here are listings in "classical"
BASIC:

```
1000 REM SUBROUTINE CHEER
1010 REM M$ IS THE USER'S NAME
1020 R=INT(RND(0)*10)+1
1030 FOR I=1 TO R
1040 READ C$
1050 NEXT I
1060 PRINT C$;", ";M$;"!!!"
1070 RESTORE
1080 RETURN


2000 REM SUBROUTINE BOO
2010 REM M$ IS THE USER'S NAME
2020 R=INT(RND(0)*10)+1
2030 FOR I=1 TO 10+R
2040 READ C$
2050 NEXT I
2060 PRINT C$;", ";M$
2070 RESTORE
2080 RETURN


4000 DATA"YES, YES, YES","RIGHT ON","YOU GOT IT","GOOD
","SWELL","NEAT"
4010 DATA"THAT'S IT","EXCELLENT","THAT'S MY ANSWER","F
INE"
4020 DATA"NO","THAT'S WRONG","THAT'S NOT RIGHT","THAT'
S TERRIBLE"
4030 DATA"THAT'S NOT MY ANSWER","I DISAGREE","THAT CAN
'T BE RIGHT",
4040 DATA"I DON'T BELIEVE IT","THAT'S IMPOSSIBLE","NO
WAY"
```

The same subroutines, in "micro" BASIC, appear as

```
370 T=RND(10):FORI3=1TOT:READX$:NEXT:PRINT 512,CHR$(31
),X$;"!!!!!":FORI7=1TO1000:NEXT:RESTORE:RETURN:DATAYES,
GOOD,GREAT,NEAT,SWELL,RIGHT ON,I LIKE THAT,"THANKS, I
REALLY NEEDED THAT",FINE,SUPER

380 T=RND(10):FORI3=1TO10+T:READX$:NEXT:PRINTCHR$(31),
```

```
X$:RESTORE:FORI7=1TO1000:NEXT:RETURN:DATANO,"NO, NO, N
O",NO WAY, THAT'S WRONG,THAT CAN'T BE RIGHT,I DISAGREE
, I DIDN'T GET THAT ANSWER,YOU MADE A MISTAKE,THERE'S
A MISTAKE SOMEWHERE,NOPE
```

In these examples, the packed code enabled me to include the entire subroutine, including a timing loop, as a logical unit under a single line number.

Microcomputer Assisted Learning. Computer Assisted Learning, CAL is associated with the use of a computer to assist the learning process in a discovery mode, as opposed to the drill and practice mode typical of CAI programming. Computer simulated experiments, populations, or environments are generated in which the student may "see what happens if..." variables are changed. As an example, a student may explore the effect of nuclear charge on the depth of the potential energy well representing a 3s orbital. Typically, the computer will show the result of changes in one or more parameters through plots (potential energy vs. distance from the nucleus as a function of Z, in this example). Quite clearly, hardware requirements for CAL programming include extensive interactive graphics capability. The CAL technique has been thoroughly discussed by the CUSC group, in Britain (8), and by Alfred Bork, who has applied it to physics at the University of California, Irvine (9).

Application of microcomputer systems to CAL necessarily depends heavily on the graphics capability of the hardware. Each microcomputer vendor appears to have independently developed a strategy for the generation of graphics. The most familiar strategy employs character block graphics, in which segments of the basic character dot matrix are selectively turned on or off. These blocks may be addressed anywhere on the video display screen. The shape of each individual graphics character is determined by the character generation hardware, and varies from vendor to vendor. This scheme is used by Commodore, Radio Shack and Heath, among the more familiar vendors. An alternate (and more flexible) approach to graphics generation provides for individual address of points on the video display. Graphic resolution is limited by the resolution of the video display, itself, rather than by re-

strictions on the size of displayable character
blocks, and by the available video display refresh
memory. Point addressing, combined with appropriate
software, permits the generation of vector based
graphics, in which the computer can be directed to
construct a line between two points at opposite
corners of the display, for example. The point addressing mode permits facile construction of curves
which resemble typical "Calcomp" plots, and which are
fully adequate for CAL (vide infra). Vector graphics
are available on the Apple II, among the popular low
cost systems.
        In chemistry, outstanding examples of interactive
graphics include the work of Stan Smith, at the
University of Illinois, with the PLATO system, Bill
Butler and Henry Griffin at the University of
Michigan, using PET computers, and of Scott Owen, at
Atlanta University, using an Apple II (1-6). Each of
these systems fully utilizes "foolproof" hardware and
software. PLATO software was specifically designed
for educational purposes. The software of the PET and
Apple II is general purpose microcomputer BASIC. The
PET computer uses character block graphics, and the
examples produced by Butler represent outstanding examples of this technique. The PLATO system originally
employed a computer driven slide projector for high
resolution graphics, but current programming makes ex-
clusive use of vector graphing, which is also used on
the Apple II. It is clear that any of the learning
techniques which depend heavily on graphics will also
be device dependent. It is also clear that any technique which can be applied <u>via</u> a large computer can
also be applied <u>via</u> an appropriate microcomputer,
often with superior technical quality.

        <u>Microcomputer Managed Instruction</u>. Computer
Managed Instruction, CMI usually is distinguished from
CAI or CAL by the interaction between the computer and
an instructor or administrator, rather than the student. Among the few CMI systems which are directly
accessed by the student, the CHEM TIPS system at the
University of Wisconsin is outstanding. In general,
grades are recorded and "massaged", and exams are generated by the instructor. CMI systems typically require large amounts of fast mass storage, and are not
easily adaptable to microcomputer systems. An inter-

esting exception has recently been developed by Art
Lepley, at Marshall University. Using a collection of
"hobbyist" cards, a floppy disk, and a letter quality
printer, Lepley has converted a low cost word processing system into an examination generator. Though his
system does not yet have the capabilities of the large
scale examination generators described by K.J.
Johnson, Ron Collins and John Moore, future software
developments will make it a close contender.

In the area of record keeping and associated
tasks, the concept of "clustering" stand alone microcomputers around a central disk storage unit suggests
that a cost effective means of sharing high speed
read/write mass storage is on the horizon. At IIT, a
Nestar Cluster One system supports eight PET and seven
Apple II computers within the IIT Educational Technology Center. Though it has been in service for only a
few months, it has already shown that it has the capability to provide educational computing support comparable with that provided through the PRIME 400 system
at a small fraction of the cost. Each of the microcomputer stations can utilize all of the special
features (graphics capability, timing, etc.) inherent
in the individual machine. Moreover, because all computation is performed at independent microcomputer
stations, the system is virtually immune to massive
breakdown. More recently, smaller (and less expensive) systems have been offered which support up to
eight PET computers, and larger systems which can
support up to 64 Apple computers with up to 40 Mbytes
of storage are now being shipped by Nestar Systems.

In summary, I believe that the real contribution
of microcomputers to computer enhanced education lies
not in the cost effectiveness, alone, but to a larger
degree in the inherent "foolproof" nature of consumer
oriented products. Expansion to applications requiring greater storage (CMI) or sophisticated graphics
(CAL) depends on the willingness of an adventuresome
educator to extend the capability of a microcomputer
to its limit (thereby encountering some of the pecu-
liarities of computers which may best be left to computer scientists) or the development of these capabilities in "foolproof" personal computers designed to
meet consumer demand. In truth, the development of
challenging software for today's microcomputers will

doubtless hasten the day when truly sophisticated personal computers reach the mass market.

References Cited

1. Moore, J.W., and Collins, R.W., J. Chem. Educ., 1979, 56, 140.
2. Lower, S.K., Gerhold, G., Smith, S.G., Johnson, K.J., and Moore, J.W., J. Chem. Educ., 1979, 56, 219.
3. Solzberg, L.J., J. Chem. Educ., 1979, 56, 644.
4. Gerhold, G., Macero, D.J., Lyndrup, M.L., and Moore, J.W., J. Chem. Educ., 1979, 56, 701.
5. Moore, J.W., Gerhold, G., Breneman, G.L. Owen, G.S., Butler, W., Smith, S.G., and Lyndrup, M.L., J. Chem. Educ., 1979, 56, 776.
6. Moore, J.W., Gerhold, G., Bishop, R.D., Gelder, J.I., Pollnow, G.F., and Owen, G.S., J. Chem. Educ. 1980, 57, 93.
7. Brubaker, G.R., Creative Computing, 1979, 5(10), 130.
8. McKenzie, J., Elton, L., and Lewis, R., Eds., Interactive Computer Graphics in Science Teaching, Ellis Horwood, Chichester, 1978.
9. Bork, A.M., Amer. J. Phys., 1975, 43, 81.

Team Design Approach for Teaching
Computer/Instrument Interfacing
in a Liberal Arts College

The use of computerized data aquisition and control systems
is rapidly becoming an integral part of many chemical instruments. Because of this, principles involved in the operation and
use of these systems are becoming increasingly important to
include in an undergraduate science curriculum. While it may be
true that the majority of chemists will not likely have to program computers in assembly language or actually build digital
interfacing, they are likely to be required to purchase computerized instruments, use them, and/or understand the advantages
and limitations of them so they can intelligently determine
appropriate applications. We have felt that the best way to ex-
pose undergraduate students to these concepts is to provide
opportunity for a hands-on experience in designing and implementing a simple on-line microcomputer or minicomputer data
aquisition or control system for use with a common chemical
instrument.

Courses dealing with interfacing computers to scientific
instruments have been offered in increasing numbers at various
graduate-level chemistry departments, undergraduate engineering
and technical schools as well as some undergraduate liberal arts
colleges over the past several years. As a result of these
offerings a number of textbooks, laboratory manuals and suggested
curricula and course content guides have become available (1-3).
Even more recently materials and workshops have become available
suggesting ways and providing equipment to incorporate microcomputers into such courses (4-6). The focus of this paper is
not to describe content or equipment as much as to describe a
method of administering the course which we have found helpful
in meeting the needs of students at a small liberal arts college.
A number of somewhat unique problems in offering such a

course at a small liberal arts college arise which offer an
additional challenge to the instructor. The problems involve:
(1) the relatively small number of students in a given department, (2) the heterogeneity of student interests and backgrounds
relating to computers, electronics, and interfacing techniques
even within a given department, (3) the time available in a
typical student schedule for dealing with computer-instrument
interfacing, (4) the expense of the equipment needed to teach
such a course and maintain the equipment, (5) the appropriate
balance of depth and breadth of topics covered, particularly if
students from several departments populate the course.

One way to make such a course interesting and applicable to
a variety of student backgrounds and interests is to use a
"team design" approach which capitalizes on the interests and
abilities of students from several undergraduate majors. We
have offered such a course on alternate years for the past six
years and have found it to be increasingly popular with students
from chemistry, physics, computer science, and business/systems
backgrounds.

Normally the course enrollment is limited to twelve students
to minimize conflicts in using the equipment available. In the
past, two computer systems have been available in the laboratory
for student use. During this current year, because our new
Information Systems Department at Taylor has recently decided to
make this course a requirement for its majors, scheduling pressures forced the enrollment to 23 students. A third computer
system has been added to help service these students. One
system includes a PDP8/L minicomputer with 8K of memory, a
Tennecomp 1371 cassette tape, a teletype and a Heath Computer
Interface Buffer to be used in conjunction with Heath's Analog
Digital Designer and associated digital logic modules. While
this system is somewhat outdated at present and will likely be
replaced with a microcomputer system in the near future, it has
proven to be very effective for use in this course and still
functions quite well. The second system is a Mini-Micro-
Designer (MMD-1/MI) with a memory interface extension from E & L
Instruments. This system is based on an 8080A chip and includes
256 words of ROM, 2½K of RAM, a cassette tape recorder and teletype. A student-designed and constructed interface permits this
system to be used as a high speed terminal on Taylor's PDP 11/70
time-sharing system. This interface is based on an AY-5-1013A
UART and was constructed in a semipermanent form on EXP 300
wireless breadboarding cards. This interface permitted the
MMD-1 to be connected to a standard 20 ma current-loop terminal
line from the DEC 11/70 in place of a TTY or DEC-Writer. An
NE555 was used as a clock oscillator the frequency of which

determined the baud rate at which the UART would operate. Although higher baud rates are possible, 300 baud proved to be fast enough for the majority of our needs. Larger programs can be written in standard 8080A assembly language using one of the available editors on the PDP 11/70 system and then cross-assembled into the appropriate binary code for the 8080A using a student-written cross assembler. The assembled program can then be dumped directly to the MMD-1/MI memory for testing. This procedure is not essential for the course but it maximizes the use of the limited 2½K of RAM on the MMD-1/MI. In addition, the design and implementation of both the interfacing module for the MMD-1/MI and the writing of the cross-assembler provided excellent projects for some of the more advanced undergraduate students. The cost of the interfacing was approximately $50. An alternate approach, which did not involve the use of the interface, consisted of doing the editing and assembling on the 11/70 and then punching a binary tape. This binary tape was then brought from the computing center to the laboratory and read into the MMD-1 via a teletype.

The third and newest system is an Apple II microcomputer with 48K of core, color T.V. monitor, real-time clock, and a single floppy disk. A standard hobby/prototype board for the Apple II was used, along with ribbon cables to bring the necessary I/O bus signals out to a wireless breadboarding socket for easy interfacing access. Extremely little modification is required to permit the same interface circuits built by students for the MMD-1/MI system projects to be used on the Apple II. The additional advantages of more memory, doing most of the I/O programming in BASIC using PEEK and POKE statements, much more convenient editing and assembling when assembly language routines are written, as well as the capabilities of color graphics make this system very attractive.

The overall objectives of the course are (1) to familiarize students with the basic software and hardware necessary to implement an on-line micro- or minicomputer system, (2) to permit students to develop a "feel" for the way a micro- or minicomputer operates internally and the ways it can be used to interact with the external world and (3) to permit students to use selected systems techniques in implementing an on-line micro- or minicomputer system. The major activities used to accomplish these objectives are:

(a) Read about and discuss the capabilities, applications and typical architecture of micro- and minicomputers.

(b) Program small-memory micro- or minicomputers in assembly language for on-line operation with a common scientific instrument.

(c) Read logic diagrams and use Boolean algebra and truth tables to design simple digital circuits.

(d) Use basic I/O programming.

(e) Read about and discuss basic hardware operation, limitations, advantages, and applications.

(f) Use selected data aquisition and reduction techniques.

(g) Use PERT and/or GANTT planning techniques.

(h) Co-ordinate group activities.

(i) Work on a team sub-system project.

(j) Employ acceptable communication, reporting and documenting techniques.

The class is divided into groups of four students near the beginning of the course and each group is assigned a specific project which must be completed by the end of the course.

Approximately the first third of the course is spent on the computer architecture and internal data flow. During this time all students learn to write, edit, assemble and debug assembly language programs for either the PDP8/L or MMD-1/MI as well as use the library of subroutines available for each system. The final program written by each student during this first third of the course involves programming the computer to print out in a specified format on the teletype the contents of a block of data stored in the computer. This program is simple but involves programming the computer to request and accept the starting address of the data block as well as the number of points to be printed out, using subroutines to handle and inter-convert binary, octal, BCD and ASCII code and print messages. The program can be used as a part of the software for the group project due at the end of the course.

The second third of the course involves an introduction to digital logic, drawing and reading logic schematics and using Boolean algebra and truth tables to design simple digital circuits. All students are required to actually wire and test several digital circuits during this time.

The last third of the course is devoted to studying I/O commands, interrupt programming, D/A and A/D conversion techniques and other concepts necessary for completing the final project. Also, during this last third, more general concepts of comparative computer architecture, data reduction techniques, and larger mini- and microcomputer system configurations are discussed. The laboratory time during this phase is devoted entirely to the groups completing their assigned projects.

The activity sequence varies somewhat each time the course is offered since the activities are designed to accommodate the backgrounds and interests of the students enrolled as much as possible as well as expedite the completion of the assigned projects. Several activities occur simultaneously at various times and therefore every student does not follow the same sequence of activities at every point in the course. This is necessary since each member of a group design team must develop certain skills and complete his own team sub-system assignment as well as meet certain minimum proficiency requirements common to all students enrolled.

In each group of four students, one student is assigned or volunteers to be the "group coordinator", one student is a "hardware specialist", another is a "software specialist" and another is an "applications specialist." Typically, the group co-ordinator might be a business/systems or computer science major interested in management skills, the software specialist frequently tends to be a computer science or physics major with strong interests in programming, the hardware specialist is any student with previous experience or interest in electronics and the applications specialist normally is a chemist or a student with some background in chemistry. Each student is given the job descriptions for the various positions before volunteering or accepting the specific assignment. A brief outline of the more detailed job descriptions given the students follows:

Project Co-ordinator
1. Plan and co-ordinate all group activities and make individual assignments to members.
2. Evaluate each group member's performance.
3. Understand and be able to explain all hardware and software functions but not necessarily write or configure all of it.
4. Write weekly progress reports.
5. Attend all project co-ordinators meetings with instructor.
6. Work on some related aspect of the project or fill-in if needed.
7. Direct the overall system design and specification writing (in conjunction with the specialists) and prepare the final system documentation report.

Software Specialist
1. Be responsible for writing software to accomplish the project goal. Design and integrate the overall software package and work out with project

manager the assignment of selected software modules
to others in the group.

Hardware Specialist
1. Design and construct all interfacing.
2. Write an assembly language program to activate
   the interfacing.
3. Test the overall system.

Applications Specialist
1. Determine data reduction and display specifications.
2. Be responsible to have the instrument working and
   the signal properly conditioned to connect to the
   hardware specialist's interfacing.
3. Work closely with the hardware specialist in the
   design, construction and testing of the overall
   system.
4. Act as group resource person for all matters per-
   taining to the application (ie., instrumental,
   theoretical and chemical).

Although each group member has tasks which capitalize on
his own unique interests and abilities, there are several activities and proficiency requirements common to all group members.
These are listed below:
1. Write, edit, assemble and successfully operate
   a program to print out in a specified format
   on the teletype the contents of a data block
   stored in the PDP8/L or MMD-1/MI.
2. Explain the operation and uses of common digital
   circuit components and sub-systems such as:
   gates, flip-flops, latches, counters, A/D
   converters, D/A converters, decoders, operational
   amplifiers, V/F converters, timing and sampling
   circuits, one-shots, shift registers, sample and
   hold circuits, etc.
3. Read and construct digital schematic circuit
   diagrams. Design a simple system (on paper) to
   solve a given problem using a truth table.
4. Understand, explain and answer questions about
   all phases of the final hardware and software
   configuration of the group project.
5. Receive at least an average of an "acceptable"
   rating on the contribution to the group
   project.

The overall project specifications and specific job descriptions are made by the instructor. These selections and assignments are key to the success or failure of the course. The overall project complexity should be appropriate for the ability level of the students in the group so the students will be challenged yet be capable of successfully implementing the system in the time allowed. The project tasks must be broken into assignments to individuals which are as independent from one another as possible so the performance of each group member is clearly demonstrated.

Two representative group projects are described briefly below. The first of these used the PDP8/L system and the second used the MMD-1/MI.

1. Gas Chromatography: Real Time Data Analysis, Display and Peak Only Storage.

    Data points from a gas chromatograph are collected and analyzed in real time using program interrupt to determine if the points are on a peak. Only points on peaks are stored in the computer memory. The contents of the memory data block are continuously displayed on an oscilloscope so the updated accumulation of data can be monitored point by point as it is collected. The sweep of the oscilloscope is triggered by the computer each time the computer cycles through the memory block and the sweep rate is adjusted by the oscilloscope controls to give a stable display. Only one D/A converter is required to display the data using this approach. After all the data has been collected, the computer jumps to a data reduction program written in FOCAL which computes adjusted peak retention times and percent composition and then prints the results out on the teletype. As a user selected option a hard copy of the raw data can also be printed out. Initial user supplied inputs supplied in response to prompters specify such parameters as the number of data points to collect, the background noise level required for the point to be considered as being on a peak or to determine the end of the peak.

2. Computer Controlled Titrator and Display of Titration Data.

    A Sargent constant-rate burette is controlled by the microcomputer to deliver titrant. The progress

of a simple acid-base titration is monitored using
an Orion 701 pH meter with BCD readout terminals
on the rear of the meter used as an input to the
microcomputer. Each data point is stored and continuously displayed on an oscilloscope. The
titration can be stopped automatically either when
a specified end point potential is attained or
when the entire titration curve has been defined.
In either case the endpoint volume is printed out
by the computer on a teletype. As an option, the
entire set of data defining the titration curve
can be printed out. As an added degree of complexity for higher ability students a "slow down"
potential can be specified which causes the buret
to add titrant drops more slowly just prior to the
endpoint. This minimizes the error due to solution
mixing and electrode response. All of the above
can be done without using floating point arithmetic. Separate student projects, not necessarily
a part of the Micro/Minicomputer Systems course
have involved implementing portions of a floating
point package in the limited memory of the MMD-1/MI
to do normality calculations from the titration
data. The initial design and construction of the
basic hardware and software subroutines was a
separate independent project for a more advanced
student. Projects in the Micro/Minicomputer class
typically center around setting up the previously
designed system, thoroughly understanding all
aspects and making assigned modifications in the
hardware or software to improve or extend the
operation.

During the most recent offering of the course, the Apple II
system was also used to implement a computer controlled titrator.
Real-time, high resolution, color graphics display of the titration curve on a T.V. while the data was being collected along
with normality calculations at the end of the titration were
part of the project. The configured system was capable of accepting user specified values of pH or millivolts to define the
endpoint as well as the point at which continuous delivery of
titrant could be slowed down to a dropwise addition. The time
between drops could also be specified by the user. APPLESOFT
BASIC makes programming these data reduction, control and display operations quite simple compared to systems like the
MMD-1/MI. These aspects of the overall project also provide good

software and systems design experience for the less hardware-oriented students in the course. In addition, and not insignificantly, we have found that the configured Apple II titrator is very useful for demonstrating titration curves for a variety of chemical systems in general chemistry and analytical chemistry courses. The interface built for this system is simple and can easily be used for interfacing the Apple II to other laboratory instruments. Work is now in progress to do this.

In summary, the "team design" approach has appealed to students because: (1) Topics studied can be immediately applied in developing a functioning system, (2) While all students are required to understand all phases of the final system their group implements, a given student can focus his efforts and time on that facet of the project with which he feels most comfortable or in which he is most interested. For this reason, the only common pre-requisite to the course is a minimum of one semester of an introductory computer science course. (3) Since the course ends with the teams demonstrating their complete functioning systems, students derive a sense of satisfaction in seeing how what they have learned leads so immediately to a useful application. (4) Students develop their ability to communicate and co-operate with other students from several major areas. (5) Projects can be chosen to be sufficiently similar to capitalize on a set of minimum proficiency requirements which must be met by all students, yet these projects can be sufficiently different that one group's design cannot satisfy the requirements for another group's design specifications. (6) Specific assignments on various projects can be made more or less sophisticated to allow each student to be challenged but yet successful. These assignments provide the greatest challenge to the instructor and are the key to the overall success of the course.

REFERENCES

1. Perone, S. and D.O. Jones. "Digital Computers in Scientific Instrumentation: Applications to Chemistry." McGraw-Hill New York, 1973.
2. Wilkins, C.L., S.P. Perone, C.E. Klopfenstein, R.C. Williams, and D.E. Jones. "Digital Electronics and Laboratory Computer Experiments." Plenum, New York, 1975.
3. COSINE Committee of the Commission on Education National Academy of Engineering. "Minicomputers in the Digital Laboratory Program." Washington, D.C. 1972.
4. Pugsley, J.H. and C.B. Silio Jr. "Microprocessors in Undergraduate Digital Design Laboratories." Computers in Education. 1:1 (1976).

5. Rony, P.R., D.G. Larsen, J.A. Titus and C.A. Titus.
        Bugbooks V - VII. E & L Instruments, Connecticut.
        (1977-78).
6. Technical Education Research Centers, 575 Technology Sq.,
        Cambridge, Massachusetts. "The Microcomputer Workshop."

B. J. MACERO, L. W. BURGESS, JR.,
T. M. BANKS, and D. CHAPMAN

CHAPTER 17

Microcomputers in Research and Teaching: An Approach to Machine Independence

ABSTRACT

The development of low-cost microcomputers gives chemists a new and powerful tool for use in research and teaching; however, this use is not yet widespread because the variety of such machines, each with its own microprocessor instruction set, hardware configuration, and bus structure, is such that it is difficult to decide which microcomputer and what software are best suited for the job in hand. At our laboratory, we have in use 8080/Z80-based microcomputer systems ranging from manufacturer-assembled and tested units to those configured from components or kits; all are outfitted with various peripheral devices such as CRT and printing terminals, floppy disk drives, video graphics, hard copy digital plotters, and analog-to-digital and digital-to-analog converters. To program these we have available BASIC, FORTRAN, PASCAL, and other, less familiar high level programming languages, in addition to assemblers and text editors specific to each system. We were able to reduce this duplication of systems' hardware and software by adopting for our microcomputers identical hardware and "software" bus structures, i.e., the S-100 hardware bus and the CP/M^{TM} disk operating system. Employing a modular programming approach in conjunction with FORTRAN, we are now able to generate, with a minimum of effort, 8080/Z80 machine code programs which will run on any of our microcomputer systems.

[†]Present address: Laser Analytic, Inc., Bedford, MA 01730

INTRODUCTION
    The ready availability of low-cost microprocessor chips and
microcomputers has spurred an interest in automating many chemical systems; several computer-controlled instruments already
have been reported.[1-3] These range from simple microcontrollers
with limited software capabilities to micro- and minicomputer-
based systems of varying degrees of complexity and flexibility.
[4-8] While minicomputers are generally faster than microcomputers and, in some cases, make available to the end user a considerable amount of software developed over the years, the cost
of microcomputer technology is approximately an order of magnitude less than that of minicomputers. This cost advantage, however, is offset by the fact that many microcomputer systems
suffer from inadequate software support, lack of mass storage
and retrieval capability, insufficient memory, and lack of
standardization. The result is that the chemist or other researcher desiring to assemble a computer-controlled system is
often faced with a formidable, and discouraging task.
    After several years of experience with microcomputer hard-
ware and software, both off the shelf and that developed by us
in our laboratory, we have evolved a design philosophy aimed at
making these decisions and acquisitions less painful and more
knowledgeable. This paper sets forth design considerations and
interfacing guidelines, readily adaptable to various types of
microcomputer usage, i.e., CAI, calculations, data reduction,
or instrument control, which are useful to keep in mind when
choosing microcomputer hardware and software.
    The most important consideration in configuring a micro-
computer system is to use readily available, standard hardware
and software components. Standardization allows one to take
advantage of the wide variety of low cost components, circuit
boards, and other digital devices which are now available for
use with microcomputers. In the case of 8-bit machines, this
is most readily accomplished by the adoption of a standard hardware bus structure, the S-100 bus, a parallel 100-pin backplane,
originally designed for 8080-type CPU systems, and a standard
"software bus" structure, CP/M[TM]*, a monitor Control Program
for Microcomputer system development.
HARDWARE CONSIDERATIONS
    The S-100 bus was introduced in 1975 by MITS, Inc. when the
Altair microcomputer kit was first brought out; its use has since
spread throughout the microcomputer industry with the result that
over 100 manufacturers now offer S-100 compatible microcomputer

---

*CP/M[TM] is a registered trademark of Digital Research, CA

boards of various types. This ubiquitous bus has become something of a de facto standard for 8-bit microcomputers; it may receive official sanction as such now that the proposal for standardizing the S-100 bus, put forth by the IEEE Microprocessor Standards Committee, has been favorably acted upon.

In the emerging realm of 16-bit microprocessor-based machines, the S-100 bus also shows promise of becoming more widespread. At present, several microcomputer manufacturers market 16-bit S-100 compatible systems, i.e., the Intel 8086-based microcomputer sold by Seattle Computer Products, Inc. 1114 Industry Drive, Seattle, WA 98188. We anticipate that this trend will continue and that more of the new 16-bit microprocessor chips will show up on S-100 compatible cpu boards, and also that the forthcoming megabyte density Winchester disk drives will be adapted for use with S-100 bus systems.

Such hardware compatibility is vitally important to the end user because it allows him to take advantage of the wide variety of low-cost, state-of-the-art microcomputer boards such as memory, disk controllers, clocks, and graphics boards, now available. The ready availability of such accessories, in turn, leads to wider and more diverse usage which further spurs the development of even newer products. The user who assembles a system around a standardized hardware configuration, also reaps the bonus of reduced down time, since "off-the-shelf" boards or components and excellent trouble shooting advice are generally available locally. A further benefit accruing to the user of standardized components is that perceived hardware needs are often underestimated, and, as soon as a person gains experience with microcomputers, new uses and tasks for his system are envisioned. Using S-100 bus compatible components permits the user to assemble a minimum configuration system commensurate with his or her initial budget and expertise while still retaining the option to easily and economically expand the system in the future.

At our laboratory, we settled upon using 8-bit microcomputer systems early;[9] however, after some frustrating experience with a single-manufacturer bus structure, we adopted the S-100 bus format. We now have three S-100 bus compatible microcomputer systems; all are "hybrid" systems, i.e., the mainframe and power supply are from one manufacturer, the cpu board from another, memory boards from several sources, and so on. This approach allows us to maintain versatility while keeping us from getting hopelessly locked in to one manufacturer's hardware.

Many of the laboratory microcomputer systems described in previous reports lack rapid, adequate mass storage and retrieval capability. Several make use of paper and cassette tape storage;

however, these are slow and cumbersome when compared with floppy disk storage. Floppy disk mass storage capability not only provides the user with random access to data files and a high rate of data transfer, e.g., 31,250 8-bit words per second, at moderate cost ($1500-2000 for a two drive system), but also provides easy system-to-system data and program transportability. In order to incorporate this important feature into a system, however, the user must standardize on diskettes formatted according to the IBM soft sector recording format. This is readily done using an S-100 bus compatible single density formatting disk controller such as the S.D. Systems (P.O. Box 28810, Dallas, TX 75228) "Versafloppy" or the Micromatron, Inc. (1620 Montgomery Street, San Francisco, CA 94111) "Doubler" a single or double density formatting disk controller.

The computer communicates with various peripheral devices such as console, a printer, or analog-to-digital and digital-to-analog converters by means of input and output, I/O, ports. Dedicated ports are often included in boards such as disk controllers; however, some type of general I/O capability with channels for both serial and parallel data transmission must be included in a system designed for experimental monitoring and control. A good example of a versatile I/O card is the Cromemco (280 Bernardo Avenue, Mountain View, CA 94040) dual universal asynchronous receiver/transmitter, TU-ART, digital interface. The TU-ART features two serial channels for duplex serial data exchange and two parallel channels, in addition to ten presettable computer-controlled timers. The serial data ports accept TTL, 20 mA or RS/232 level signals, while the parallel ports have provision for software address reverse and interrupt sense lines. The TU-ART also offers the user sophisticated interrupt handling capabilities such as on-board priority encoding, interrupt generation and acknowledgement, and daisy chain expandability.

Our system console consists of a Sanyo Electric, Inc. (Compton, CA) Model 4209 Video Monitor which is interfaced to the computer with a Dynabyte Model 57 Naked Terminal board. The Naked Terminal contains two switch selectable, dedicated I/O ports for communication with the S-100 bus host system and its own on board memory for the video display buffer. The display consists of 24 lines of up to 80 characters per line with each character composed from a 5 x 7 dot matrix. Input to the terminal is from a full ASCII encoded keyboard, Model 756 (George Risk, Inc. Kimball, NB). A block diagram of the microcomputer system is shown in Figure 1.

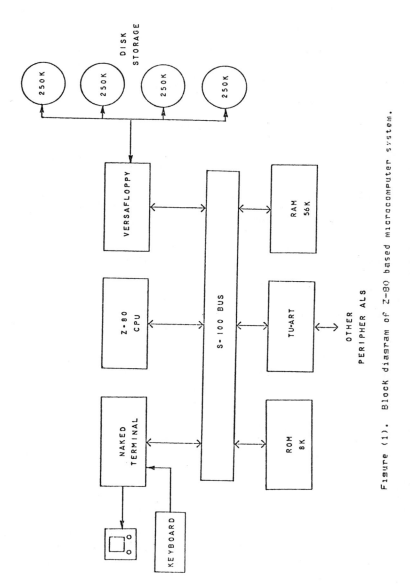

Figure (1). Block diagram of Z-80 based microcomputer system.

211

SOFTWARE CONSIDERATIONS
The cost of hardware is limited. Only a finite number of
elements are needed to assemble a powerful, versatile microcomputer system. Software needs, however, are virtually infinite, and in the long run, the lion's share of the cost of operating a computer system will be that of software development.
It makes sense, then, to choose a microcomputer system with the
more versatile, more powerful, easier-to-use software than one
with a faster cpu, bigger instruction set, and larger memory
size.
The actual microprocessor chip found on the cpu card is
unimportant if one is to program only in machine, in assembly
code, or in BASIC. All mainframe vendors supply a version of
these programming aids to support the particular chip employed
in their systems; however, our aim was to achieve machine independence and this goal dictated that we choose a cpu based on
the 8080 processor since the widespread use of this chip has
resulted in the development of a large reservoir of software
applications. A counterpart of the 8080 is the Z-80, another
8-bit microprocessor which is able to execute all the 8080 instructions as well as an extra 80 instructions of its own, in
addition to having a more powerful interrupt structure and the
ability to generate very efficient code for use in time-critical
applications. By choosing the Z-80 as our system cpu, we were
able to avail ourselves of all the software developed for the
8080 and still have access to the additional programming power
of the Z-80.

THE CP/M^{TM} OPERATING SYSTEM
The software required to manage the hardware and logical
resources of a computer system is the operating system. The
simplest operating system consists of a keyboard monitor, i.e.,
a small set of executive commands which enable the user to communicate with and to exercise control over the computer from a
CRT or a TTY terminal. For more sophisticated hardware configurations such as one featuring floppy disk drives and a printer
in addition to a console, a general purpose disk based operating
system is desirable.
Most microcomputer manufacturers provide disk operating
software for use with their hardware configurations; however,
these are usually specific to the manufacturer's system or
systems. We wanted a more universal and standard operating
system; one which would provide a high degree of program transportability, and which could take advantage of the wide variety
of software development tools such as text editors, assemblers,
and debugging software, now available for use with microcomputers. These requirements led us to the selection of CP/M^{TM}, a

monitor Control Program for Microcomputer system development
(Digital Research, P.O. Box 579, Pacific Grove, CA 93950) as our
microcomputer disk operating system.

In 1973, Version 1.0 of CP/M was developed by G. Kildall of
Digital Research as an operating system for use with 8080-based
computers. In 1976, this was upgraded to Version 1.3 to provide
program transportability between different 8080 machines using
standard 8-inch IBM formatted diskettes.[10] The overwhelming response to this operating system led to the development of CP/M
configurations (now Version 1.4) which provided operating systems for most every type of 8080 or Z80 microcomputer system
employing either 5½-inch (minifloppy) or 8-inch floppy disk
drives capable of recording in either single or double density,
single or double sided, or in hard or soft sectored formats.
More importantly, CP/M can be easily altered in the field to
operate with a non-standard or "hybrid" system, i.e., one with
function boards from several different sources. The latest version of CP/M, Version 2.0, is able to handle up to 16 disk drives,
each with up to 8 megabytes of memory, and thus can serve as an
operating system for high density, high capacity Winchester and
other hard disk drives.

Another operating system, also developed by Digital Research
of California, for use with 8080-, Z80-, and 8085-based microcomputers that can be used with CP/M software is MP/M[TM]*. This
is a multi-terminal operating system which can support real-time
multiprogramming at each of several terminals connected to one
microcomputer. MP/M uses the file structures of CP/M 2.0, requires a minimum of approximately 16K bytes of memory, and can
simultaneously run editors, program translators, background
printing spoolers, or can be used for either data entry or data-
base access from remote terminals. These last two features are
especially useful in interconnecting several microcomputers in
a research or industrial laboratory for use in a hierarchichal
mode.

CP/M is divided into four units: the Basic machine language Input/Output System (BIOS), which provides the I/O driver
primitives for all standard user defined peripherals; the Basic
Disk Operating System (BDOS), which contains the disk file management routines; the Console Command Processor (CCP), which
is the interface between the operating system and the user console; and the Transient Program Area (TPA), i.e., all the computer's random access memory (RAM) below the operating system
and above 100H (hexidecimal radix), where various non-resident

---

*MP/M is a trademark of Digital Research, CA

operating system commands and all user-orignated programs are executed.

With the exception of the BIOS, all of these are hardware independent. It was necessary, therefore, to customize the CP/M BIOS to operate with the specific hardware making up our Z-80-based microcomputer. The general procedure used to adapt CP/M to any system is well documented by Digital Research in their System Alteration Guide; however, in our case, this task was facilitated by our choice of disk controller, i.e., the S.D. System "Versafloppy", which comes with CP/M compatible control and diagnostic software encoded on two 2708 EPROMS. The procedure is described below.

CUSTOMIZING CP/M TO A "HYBRID" SYSTEM

When the system power is turned on, the CPU hardware forces a jump to a memory location selected by an on board address select switch. This jump is then directed to a Cromemco Z-80 monitor program,[11] stored in a PROM located on a Cromemco "Bytesaver II" PROM board residing at the top 8K of memory from location E000 to FFFFH. The Cromemco keyboard monitor software contains executive commands for examining, changing, moving, and comparing memory, altering CPU register contents, controlling serial and parallel ports, and programming PROMS.

With the keyboard monitor, then, the user can bring up the system and begin the process of customizing CP/M. To do this, the Versafloppy BIOS PROM must be altered to make it compatible with the Shugart 801 disk drives and the specific console hardware. Only a few bytes of this 1K byte BIOS program need be changed for this purpose: a table of variables which selects operation on full size (8-inch) disk drives, a delay constant for the CPU clock, and the status and data port designations and masks selected for use with our video terminal. Once these changes have been made, a new PROM is written using the Byte-savers PROM programming capability, then the Versafloppy diagnostic routines are executed to verify that the newly written BIOS routine works properly.

The next step is the first level generation of CP/M. This requires loading the distribution version of CP/M, a 16K RAM system, and modifying it to conform to the user's hardware. Normally one would have to write the assembly language routines to load the distribution version of CP/M into memory; however, with the aid of the Versafloppy disk controller diagnostic software, the initial 16K CP/M system is brought into memory by executing a load of 2EH consecutive sectors starting at track 0 sector 1. At this point, the user has in memory a rudimentary CP/M system with a BIOS residing at location 3E00H and a boot loader residing at 2880H; however, by reentering the keyboard monitor and using the monitor move commands, the various 16K CP/M routines may be relocated in memory and eventually written to disk.

Next, the jump table located between 3E00H and 3E2CH is redirected to the corresponding locations in the Versafloppy BIOS from locations F000H through F02CH; all of the Versafloppy jump table subroutine addresses may be substituted for the corresponding routines in the CP/M BIOS except those for BOOT and WBOOT. These two boot subroutines must be loaded from disk in order to be able to reconfigure the 16K CP/M system to operate with different memory sizes without having to make additional modifications to the Versafloppy PROM. The simplest way to handle the routines is to vector both jumps to a routine called GOCPM which initializes page 0 of memory and then jumps into the CCP. A power-on boot routine on the Versafloppy PROM loads the first sector on track 0 of drive A diskette into memory at 80H and then executes the routine. The rest of tracks 0 and 1 is copied into the proper memory location for the system size desired. Once the system is loaded, the program jumps to the BOOT subroutine and the CP/M disk operating system comes up. The user now has a customized version of CP/M in memory and needs only to write it back onto the first two tracks of the system diskette using the diagnostic routines, still resident at 100H. An absolute sector save of 2EH sectors from memory onto tracks 0 and 1 of the disk is done and the monitor reentered via RESET. When the Versafloppy power-on routine is executed, the CP/M monitor will boot up. Once this version of CP/M is operational, the user has a 16K operating system with all the facilities of the CP/M assembler, debugger, and editor at his disposal to expand the disk monitor to include all the available memory, and to further refine the system. In connection with the foregoing, note that S.D. Systems now offers along with the Versafloppy, a set of two diskettes: an 8-inch and a 5 ¼-inch disk, on each is a CP/M operating system configured to use the Versafloppy. With these, it is only necessary to adapt the BIOS for the specific hardware on hand.

SOME FEATURES OF CP/M

After the system is signed on, CP/M automatically logs in disk A, displays the prompt "A>", and waits for a command. Two levels of command exist: built-in commands which are part of the CCP, and transient commands which are loaded into the TPA from disk and executed. The built-in commands are:

```
     ERA  -  erase a specified file
     DIR  -  list file names in the directory
     REN  -  rename a file
     SAVE -  save the contents of memory in a file
     TYPE -  type the contents of a file
```
The colon (:) is used to log in any one of up to four disks, e.g.,
A>B: switches console operation to disk B. Similarly, files
on disk B can be accessed from disk A by preceding the file name
with B:. Thus A>TYPE B:PROG1.ASM types out a listing of the
assembly language file PROG1 which resides on disk B.
     Transient commands available with the CP/M distribution
disk are:
```
     STAT   -  list the number of bytes of storage remaining
               on the currently logged disk
      ASM   -  load the CP/M assembler and assemble a
               designated program from disk
      DDT   -  load the debugger and start execution
      PIP   -  load the Peripheral Interchange Program for
               disk file and peripheral transfer operations
       ED   -  load and execute the CP/M text editor
               program
   SYSGEN   -  create a new CP/M system diskette
```
The user can of course create his own commands or adapt those
developed by others. A number of such commands along with many
useful programs can be found on disks distributed by the CP/M
Users' Group (2248 Broadway, NY, NY 10024).
     The STAT command also allows control over the physical to
logical device assignment. There are four logical devices: CON:
(console); RDR: (paper tape reader); PUN: (paper tape punch);
LST: (output list device). The actual devices extant in any
given computer system are driven by subroutines in the BIOS, and
can be one of a number of devices, e.g., a high speed reader,
printer, cathode ray tube, cassette tape, or a user-defined
peripheral.
     The Peripheral Interchange Program, PIP, provides for oper-
ations which will load, print, punch, copy, and combine disk
files. Physical device names can be used in PIP commands which
allow the user to send nulls to the device, send an end-of-file
signal to the destination device, input and output data, and
list a file on the list device and number each line. The user
can also specify one or more PIP parameters during a copy or
list operation to transfer data in block mode, delete characters
to truncate long lines, echo all transfer operations to the console, translate upper case to lower and vice versa, copy a portion of a file starting and ending with specific characters,
and verify that a file has been copied correctly. Other pro-

grams that we have obtained for use with CP/M are: DESPOOL, a
background printing facility; and TEX, a text formatting program.

MICROCOMPUTER LANGUAGES

Programming in machine or assembly language is tedious,
time-consuming, and not very conducive to the generation of large
amounts of error-free code. The ability to develop software in
a more "friendly" language is essential. Fortunately, a number
of high-level languages are now available for use with microcomputers, most of which can be used with CP/M.

The most widely used language in the microcomputer world is
BASIC, some form of which is offered by virtually every software
and hardware microcomputer vendor; however, in the last two years
FORTRAN, PASCAL, and more recently, APL have become available.
A most attractive feature of these high-level languages for someone wishing to achieve machine independence is that they can now
be obtained on disks ready to run under CP/M.

BASIC and APL are interpreter type languages, that is, they
do not translate source code directly into machine code for execution, but rather interpret and execute the program one line
at a time. PASCAL compiles the source code into an intermediate
pseudo-code (p-code) which is in turn interpreted and then executed. FORTRAN, on the other hand, is a true compiled language;
source code is translated directly into the microcomputer machine
code. Interpreter type languages run much slower than compiler
type languages; however, error correction and program debugging
is easier with interpreter types. It is difficult to choose
a language solely on the basis of whether it is a compiled or
interpreter type. More important in the selction of a high-
level language are: the ability to write structured programs;
the ease of learning the language; the availability of a library
of software routines; and the user's experience with the language.

At our laboratory, we have CP/M compatible versions of all
these: BASIC, both interpreter and compiler versions (Microsoft
BASIC, Version 5.0, Microsoft, Inc., Bellevue, WA), PASCAL (UCSD
Version 2) and FORTRAN (also from Microsoft). The latest CP/M
compatible versions of these languages can be obtained from
Lifeboat Associates, 2248 Broadway, N.Y., N.Y. 10024. We prefer
to program in FORTRAN not only because it is faster but also
because FORTRAN is the most widely used language in scientific
applications with the result that a large body of well-documented,
scientifically-oriented FORTRAN programs and algorithms have been
developed over the years, and standardized versions of FORTRAN
are now available for microcomputer software programming, i.e.,

Microsoft FORTRAN-80. This version includes all of ANSI standard FORTRANS X3.9-1966 with the exception of the ability to handle complex quantities. Because of this standardization we experience little trouble in adapting existing FORTRAN programs for use in microcomputer applications.

The associated utility programs provided with this version give the user complete FORTRAN and assembly language program development and library management capability. In addition, the ability to link subroutines written in assembly code (as well as compiled BASIC routines) to FORTRAN-80 programs makes this package particularly well-suited for developing software in experiment control applications. As a result, the user can take advantage of the extra speed and efficiency of assembly code in time critical applications, while writing the rest of the program in an easier-to-use high level language. The combination of a high level programming language such as FORTRAN and a versatile operating system such as CP/M, provides the researcher with most of the software tools necessary to implement experiment control and data analysis.

In summary, standardizing our systems on the S-100 hardware bus and the CP/M^{TM} operating system has made us essentially machine independent; however, CP/M must be considered more important in achieving hardware independence than the S-100 bus since most microcomputer vendors are now adapting their systems to operate under CP/M, e.g., Radio Shack, Northstar. At present, we are developing a microcomputer-controlled titration system for use in our Instrumental Methods laboratory, the software for this system is being written in CP/M-compatible FORTRAN. We are using a modular top-down structured approach to the problem of organizing the experimental data acquisition and data reduction routines. When completed, our disk source files can be transported to another user who can directly load them on his S-100 CP/M-based system, and then, after a minimum of modifications such as adapting the ADC drivers and making the port assignments necessary for operation with his particular apparatus, can compile the program and run the experiment.

ACKNOWLEDGEMENT

The author wishes to thank the National Science Foundation Instructional Scientific Equipment Program and the Camille and Henry Dreyfus Foundation for their generous support of this work in the form of grants awarded to one of us [DJM].

REFERENCES

[1] N. Busch, P. Freyer, H. Szameit, Anal. Chem., 50, 2166 (1978).
[2] D. J. Leggett, Anal. Chem., 50 718 (1978).
[3] C. R. Martin, H. Freiser, Anal. Chem., 51, 803 (1979).
[4] S. Yamaguchi, T. Kusuyama, Z. Anal. Chem., 295 256 (1979).
[5] A. H. B. Wu, H. V. Malmstadt, Anal. Chem., 50, 2090 (1978).
[6] L. M. Doane, J. T. Stock, J. D. Stuart, J. Chem. Ed., 56,
[7] J. M. Aariano, W. F. Gutknecht, Anal. Chem., 48, 281 (1976).
[8] J. Slanina, F. Bakker, J. J. Mols, J. E. Ordelman, A. G. M. Bruyn-Hes, Anal. Chimica Acta, 112, 45 (1979).
[9] G. Morrow, H. Fullmer, Dr. Dobb's Journal, 3, 40 (1978).
[10] D. J. Macero, L. W. Burgess, Jr., T. M. Banks, F. C. McElroy, PROCEEDINGS, Symposium on Microcomputer-based Instrumentation, National Bureau of Standards, Gaithersburg, MD, IEEE, New York (1979), pp. 45-57.
[11] D. Siegel, Popular Electronics, 16, 67 (1979).
[12] "Cromemco Z80 Monitor Instruction Manual", Cromemco, Mountain View, CA (1979).

Use of 8008-Based Minicomputer as an
Intelligent Remote Terminal

The Datapoint 2200 is a minicomputer with an 8008-like
processor. The basic configuration is shown in Fig. 1. It is a
business processor, meant for communications applications. Its
excellent design for this purpose, low cost, and the availability
of a software floating-point emulator were important considerations in its selection. Our goal was to set up a scientific
computing laboratory as part of undergraduate teaching programs
in chemistry, engineering science and computer science, illustrating the function and use of interfaces, the principles of
computer operation and the interaction of small computers with
large remote mainframes. The design and development of interfaces and software was seen as a source of projects for students
in the three departments.

Since the funds originally available would not accommodate
a printer, all hardcopy output is produced on the remote computer. This proved to be adequate for a number of years. The
decision was made early to write all programs in assembler. This
was partly to obtain compact, efficient code; but also it is the
best way of teaching students how computers actually operate,
whether or not they ever use assembler language in their professional work. The relative simplicity of the 8008 instruction
set is an advantage in teaching undergraduate chemistry students
with a minimal computing background.

Since then the 2200 has been interfaced to a number of
laboratory instruments. A general-purpose interface was built,
having eight channels of A/D conversion, digital passthrough and
real-time programmable clock with slaved frequency counter.
Recently, a peripheral calculator based on the Motorola MM57109
microprocessor has been built to reduce dependence on the soft-

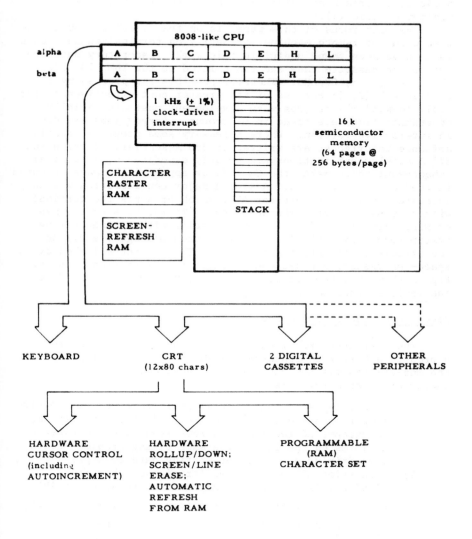

Fig. 1.   Datapoint 2200 Minicomputer

221

ware floating point computational facility. A self-contained
prototyping unit designed as a computer peripheral allows
students to breadboard digital circuits and interfaces in a
minimum of time (1).
      The main focus of this paper is on the aspect of communica-
tion with the remote machine, primarily on terminal enhancements
which have been found useful in this work. Laboratory data are
written on cassette tape and transmitted to the IBM 370 under
the host system, CP/CMS. Whenever possible the processing is
done in APL (2, 3) because of its greater cost effectiveness for
small-to-medium size jobs when development costs (especially
programmer time) are taken into account. Larger jobs are done
in interactive Fortran. Where necessary some programs, mostly
software interfaces, are written in 370 Assembler. Fortran and
370 Assembler applications are greatly expedited by a terminal
enhancement which makes the CMS editor appear to be page-oriented
instead of line-oriented. A second major consideration was an
adequate level of support for APL, which presents the terminal
with an unusually rich character set. Third, we sought a means
of minimizing the effect of communicating with a time-sharing
system (poor response time, unanticipated messages, etc.) on
the efficiency of data transmission. This is accomplished by a
space-compressed, blocked mode ("burst mode") of transmission.
Fig. 2 shows how the available 16k of memory are allocated to
the various functions of this program.

Received-Character Buffering

      The characters received from the transmission line are
stored in a buffer in space-compressed form and displayed after
the end-of-transmission character is received. There are a
number of reasons for this:

   1)   Processing each character as it is received may require
        longer than the time between characters; for example,
        to execute LF requires a screen rollup, which takes up
        to 18 ms on the 2200. The same timing problem occurs
        when driving a printer or writing data to tape.
   2)   Processing APL overstrikes (triples of characters,
        c_1 BS c_2) is time consuming; on some versions of APL
        the backspaces are clustered (e.g., c_1 d_1 e_1 BS BS BS
        c_2 d_2 e_2 instead of c_1 BS c_2 d_1 BS d_2 e_1 BS e_2) and
        arbitrarily long strings must be available in memory.
   3)   Having the entire transmission in memory allows local
        "paging" (displaying one screenful at a time) since
        many conversational systems do not offer a paging
        feature. The user can scroll down the transmitted text

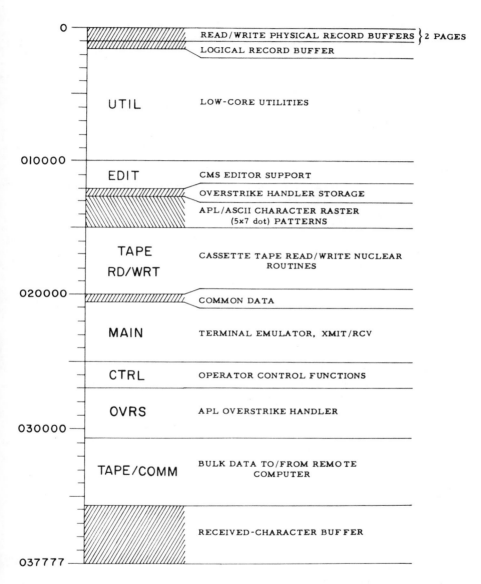

Fig. 2.   Memory Allocation for Smart Terminal Emulator

223

and optionally restart the scan at the top.

4)   A windowing feature (left and/or right truncation of each line) is useful in reading assembler listings or other lines in excess of 80 columns. When combined with buffer rescan, the user can scan the same text with different windows.

With space compression, one can store about six average (40-character) lines, or about 1/2 screen page per memory page. Ten memory pages are available after the program is loaded.

An alternative approach which is commonly used to solve the timing problem would be an interrupt-driven algorithm with a "circular" buffer, adding received characters to the "right" end of the buffer in foreground mode and processing them one at a time from the "left" in background mode. Any peripherals (printer, tape, etc.) are also driven in background mode. This would be the algorithm of choice for a machine with a small memory. This algorithm will fail at high baud rates, especially with several peripherals enabled. The 2200 can receive from a 9600 baud line only by using a tight read/store loop.

Fig. 3 shows an example of buffer processing.

APL Support

If the terminal is to handle APL, there are several special requirements:

1)   A special character set, of which more than 30 are not in the standard ASCII character set; many of those which are common to both have different binary representations.

2)   The ability to generate overstrikes, i.e., special characters formed by backspacing and overtyping one character on another.

The Datapoint 2200 has a RAM controlled character generator which can handle 128 different 7-bit codes. The RAM locations have addresses which correspond to the binary representations of the characters. Locations 0 ... 31 correspond to control characters which need not be displayed. These can be assigned to overstrikes, but there are far more than 32 which are commonly used. The underscored alphabetics alone make up twenty-six. Thus, it was necessary to devise an algorithm to allocate the 32 available "slots" to whatever overstrikes are needed at a given time. When a given overstrike rolls off the screen and does not exist anywhere below, that slot is deallocated and made available for the next "new" one. Any number of different overstrikes can be

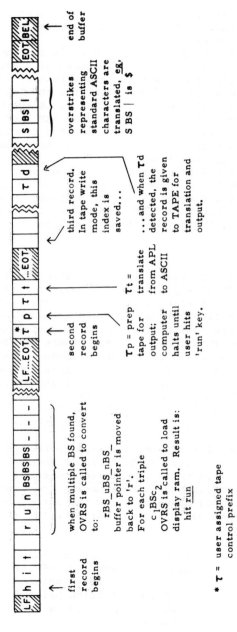

Fig. 3. Buffer Scan Example

The following describes the buffer contents shown in the figure:

first record begins (LF h i t | r u n BS BS BS - - -)

when multiple BS found, OVRS is called to convert to:

rBS_uBS_nBS_ buffer pointer is moved back to 'r'.
For each triple c_iBSc_2
OVRS is called to load display ram. Result is:
hit run

* τ = user assigned tape control prefix

(LF...EOT *τ p τ t | ...EOT)

second record begins

τp = prep tape for output; computer halts until user hits 'run' key.

τt = translate from APL to ASCII

(τ d)

third record. In tape write mode, this index is saved...

...and when τd detected, the record is given to TAPE for translation and output.

(s BS |)

overstrikes representing standard ASCII characters are translated, eg. S BS | is $

(EOT BEL)

end of buffer

225

displayed, but only 32 different ones may appear on the screen at any one time. (Any of the 32 may be repeated on the screen any number of times, however.)

The algorithm is as follows. DISP is a 32-byte list which models the RAM and the screen. The index of a byte in DISP corresponds to one of the RAM addresses (binary ASCII character codes), and each element of DISP contains the lowest screen line on which that character exists. (Line 11 is the bottom line.) A value of -1 means that the corresponding RAM slot is available. Two associated lists, CH1 and CH2, save the two individual characters for comparison with overstrikes as they come along from the keyboard or the com line.

Each time a rollup occurs, positive elements of DISP are decremented. When the value goes to -1, the corresponding line has rolled off the screen. Whenever a given overstrike is encountered, its element of DISP is assigned the value of 11 since it must be, at that time, on the bottom line.

Edit Support

Perhaps the most useful terminal enhancement is the Edit Support. This in effect converts the system line-oriented editor (4) into a screen-oriented one in which eleven lines of text are always displayed (the bottom line is a command-entry line). Modified lines are redisplayed on the correct screen line so that their contextual relationship to the others is preserved.

The text resides on the remote computer, and all the actual editing takes place there. The user enters commands which sometimes are passed on to the remote editor and sometimes are expanded into multiple commands. Not every feature of the CMS editor is available in "edit mode", but these features may be used by exiting from edit mode and using the terminal in its native or "dumb" mode.

A screen pointer (>) is displayed to the left of the current line. It can be moved up or down; if it goes offscreen another line is fetched from the file with screen rollup or rolldown. Scrolling is accomplished by instructing the remote editor to display eleven lines above or below the current screen limits.

The user thus has a virtual editor with the following instruction set:

t	display top page of file
b	display bottom page of file
P	display page above pointed line
p	display page below pointed line
D	delete pointed line & move pointer up one line

d	delete pointed line & move pointer down one line
s	scratch lines from pointed line to end of page
i	insert below pointed line
u< n >	move pointer up <n> lines
n< n >	move pointer down <n> lines
o	overlay pointed line with typed text
/c	search for character string 'c' and display page above it
f c	find character string 'c' at beginning of line and display the page above it
c/old/new	change character string 'old' to 'new' in the pointed line

The operation of this algorithm is illustrated by Fig. 4. Only the first character of each new input line is examined before it is transmitted. After the exit point labelled RETURN, the user continues to type the command and these characters are transmitted and displayed. Insert mode is enabled by typing the initial character 'i'; subsequently the only initial character which causes special action is EN, designating a null line which terminates input. After each insert line is finished, the line pointer is moved down for the next one. After the terminating null line, the program then waits for the communications line to turn around and sends 'n' (for 'next line') so that the line following the last inserted line becomes current.

In the native editor, the overlay command has the format

o < text>

so that <text> is always two columns out of register with and at least one line below the line being overlaid. In the enhanced mode the cursor is spaced over the actual text to be changed.

Cassette Tape Support

The most significant facility of this program is the exchange of data between the IBM 370 and the tape cassettes. Three routines are involved: TAPRD and TAPWRT (nuclear tape support routines) and TAPE/COMM which handles the communications aspect of the tape support. Fig. 5 shows the various functions of these routines.

Regardless of the baud rate, the major factor determining the effective transmission rate on half-duplex lines is the time required for the remote computer to "turn the line around" and request another record. Under such circumstances it is best to transmit "bursts" of as many characters as the computer will accept. On most time sharing systems, this is at least 130 and

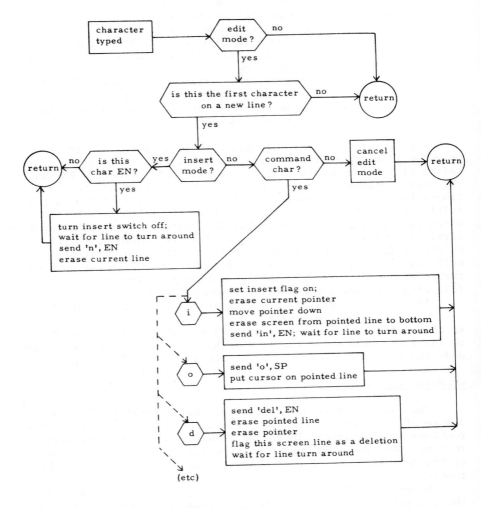

Fig. 4. Examples of Edit Support Algorithm

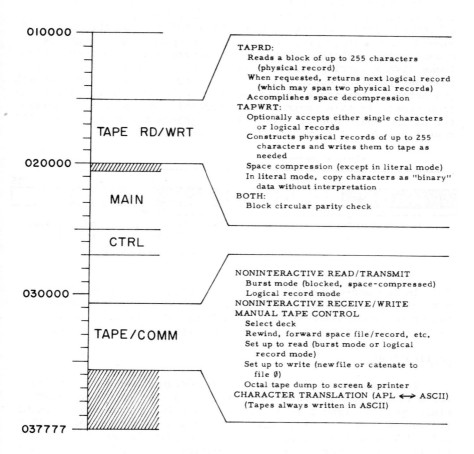

```
010000 ─────
         ─
         ─
         ─
         ─         TAPE  RD/WRT
         ─
020000 ─────  ▨▨▨▨▨▨▨▨▨▨
         ─
         ─         MAIN
         ─
         ─         CTRL
         ─
030000 ─────
         ─
         ─         TAPE/COMM
         ─
         ─  ▨▨▨▨▨▨
         ─  ▨▨▨▨▨▨
037777 ─────  ▨▨▨▨▨▨
```

TAPRD:
  Reads a block of up to 255 characters
    (physical record)
  When requested, returns next logical record
    (which may span two physical records)
  Accomplishes space decompression
TAPWRT:
  Optionally accepts either single characters
    or logical records
  Constructs physical records of up to 255
    characters and writes them to tape as
    needed
  Space compression (except in literal mode)
  In literal mode, copy characters as "binary"
    data without interpretation
BOTH:
  Block circular parity check

NONINTERACTIVE READ/TRANSMIT
  Burst mode (blocked, space-compressed)
  Logical record mode
NONINTERACTIVE RECEIVE/WRITE
MANUAL TAPE CONTROL
  Select deck
  Rewind, forward space file/record, etc.
  Set up to read (burst mode or logical
    record mode)
  Set up to write (new file or catenate to
    file Ø)
  Octal tape dump to screen & printer
CHARACTER TRANSLATION (APL ⟷ ASCII)
  (Tapes always written in ASCII)

Fig. 5.   Tape Support Routines

229

can, with the appropriate assembler interface, be arbitrarily
large on CMS. An arbitrary control character is substituted for
the EN which terminates each logical record.
   The rate can be further enhanced by space compression. A
typical scheme would be to replace a string of blanks with an
arbitrary control character (say, 011) followed by the number of
blanks. An obvious problem is that the space count cannot be
binary, since thirteen blanks would generate a space count which
looks like a carriage return, and so on. The space count must
thus be symbolic, e.g., 011, '1', '3'; but this requires a fourth
character to terminate the variable space count field. The solution chosen in this program is to transmit an alphabetic character as the space count: 011, 'B' represents two blanks, 011, 'Z'
represents 26. A count of 30 would be 011, 'Z', 011, 'D'.

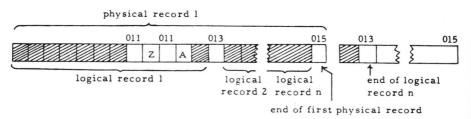

In the above example logical record 1 has 7 nonblank
characters, 27 blanks, and another nonblank character; hence
the space compression is 011, 'Z', 011, 'A'. Logical record n
is split between two physical records (transmission bursts).

Applications

   The Datapoint 2200 receives considerable use as a terminal
for a remote IBM 370 when it is not being used as a laboratory
minicomputer. (There is nothing about this program which limits
its use to IBM systems.) Considerable use is made of APL, BASIC
and FORTRAN on the remote machine. The Edit Support feature of
the terminal is very useful in preparing BASIC and FORTRAN programs for teaching and research, and in editing data. The APL
Support is absolutely essential if APL programs are to be used.
   A major application of the 2200 is for graphics. A package
of IBM 370 APL programs has been written for graphing data and
for drawing molecular structures. The output of these programs
is graphics commands for the Tektronix 4013 graphics terminal.
These may be displayed on a 4013 or they may be transmitted to
the 2200 and stored on cassette tape. Then, using a separate

2200 program, they are converted into commands for a TSP-12 pen plotter for hard copy.

The program which drives the TSP plotter is an example of the utility of a micro- or minicomputer. The plotter is designed to operate in series with a time sharing terminal, accepting graphic control characters at 134.5 or 300 baud. However, it is ineffective in this mode of operation. Since there is no internal intelligence, it cannot buffer characters in the event that a pen operation is not completed within one character-time. There is also a chronic problem of linearity on long line segments due to unequal response rate of the x and y amplifiers. The 2200 provides the needed intelligence. The nonlinearity is solved by interpolating lines above a certain length, to present shorter line segments to the plotter.

The Cassette Tape Support is useful in obtaining listings of 2200 programs using the high-speed printers on the IBM 370. We modified the 2200 Assembler to write the listing on tape so that it could be transmitted to the 370 in the burst mode described above. This is considerably faster than using a 30-character-per-second local printer, and saves the expense of purchasing and maintaining the printer. We also use this method to transmit x-ray diffraction data collected by the 2200 in local mode. These can be processed in APL or FORTRAN, graphed and returned to the 2200 for actual hardcopy production.

Conclusions

Our original goal was to establish an intimacy of communication that would allow the remote computer and its peripherals to appear to be "peripherals" under the control of the minicomputer. This is still somewhat attractive, since the cost of peripherals has not maintained the dramatic downward pace shown by computers themselves. For example, an "open line" to APL would make a superb Arithmetic Logic Unit for the 2200: by transmitting the character string

$$V \boxed{\div} M$$

one could receive the solutions of the simultaneous equations represented by M, the matrix of coefficients and V, the vector of right-hand-sides, previously "stored" in the APL workspace. However, time sharing systems are not amenable to noninteractive communications, and we have abandoned this approach. Instead, an interface for a NS MM57109 RPN calculator chip has been built, which frees the 2200 from the need for memory-consuming floating point software.

We entered the game relatively early. As a result we found
a great lack of utility software, such as program-callable tape
I/O routines, memory dump routines, etc. These are facilities
that one might take for granted from a scientific computer vendor, but such systems are far more expensive than the simple
microprocessor configurations. The same problems are likely to
occur whenever a laboratory purchases a new or unusual microprocessor, for which little software may be available.

Our department has benefited greatly from cooperation with
other departments in the University. The implementation of this
project would have been far more difficult if we had not had a
good "mix" of students from computer science and from engineering. These departments are usually looking for challenging
projects for their students.

The value of the effort invested in software soon exceeds
the value of the computer, and <u>longevity</u> should be a major consideration in the purchase of a system. The Datapoint 2200 is
an example of the superiority of minicomputers over micros.
While the power of the CPU and the memory size are quite limited,
the hardware CRT Support, and integrated high-speed digital
cassettes have given the 2200 a useful life far exceeding some
later model microprocessor-based projects that have come and
gone in the meantime. It is significant that Datapoint Corporation still markets the 2200.

Acknowledgements

This project was initiated by grant GY 11020 of the National
Science Foundation and was further funded by a Cottrell College
Science Grant and by Trinity University. Students who made major
contributions to hardware and software were: Don Kinard, Ed Van
Petten, Jr., John E. Smothers III, Richard Davis, and Michael
Millner. Much assistance with the design of hardware was ren-
dered by the Trinity University Computer Center, especially Bob
Edge, Fred Rodgers and Daniel Laser.

Literature Cited

1. Schilling, J. W., J. Chem Ed., (1979) A104.
2. Falkoff, A. D., and Iverson, K. E., Communications in APL
    Systems, IBM Technical Report 320-3022 (1973).
3. Gilman, L., and Rose, A. J., "APL: An Interactive
    Approach", John Wiley, 1974.
4. "CMS User's Guide", IBM Document #GC20-1819.

W. S. WOODWARD   CHAPTER 19

Toward an Omnivorous Interactive
Laboratory Data System

The design of interactive laboratory data systems may be
stated most effectively in terms of the identifiable interfaces
within these systems across which the flow of data and control
must be managed. This is true to some degree of course of any
sort of data processing system. The extreme variety of domains
of information and control occurring in instrumentation-oriented
laboratory data systems, however, causes the requirements of
interface design to dominate every aspect of implementation.
This paper presents a data system design procedure based upon
interface analysis. Also exhibited in some detail is a mature
general purpose microcomputer instrumentation data system design
based upon it developed at the University of North Carolina
Chemistry Department.
The functioning of a general purpose computer based
instrumentation data system involves the transfer of information
and control across as many as six interfaces:
  I.   Instrumentation/Instrumentation Controller
  II. Instrumentation Controller/Local Data Processor
  III. Human Operator/Local Processor
  IV. Mass Storage Medium/Local Processor
  V.   Remote Computational Resource/Local Processor
  VI. Application Programmer/Local Processor
Not all of these interfaces will be exhibited by every
laboratory data system application all the time. The generality
of a data system design is, however, determined by its ability
to manage efficiently each of these categories of interactions
with its environment.

233

I. Instrumentation/Instrumentation Controller
    The control of laboratory instrumentation and acquisition
of experimental data may embrace three major categories of
function:
    A. Analog/Digital/Analog Conversion:
    Although an increasing fraction of laboratory instrumenta-
tion incorporates vendor supplied data conversion circuitry and
therefore presents a digital interface to a connected data
system, this organization is far from ubiquitous. Even in the
case of instrumentation provided with internal data conversion,
system performance may be enhanced by superior conversion
circuitry external to the vendor-supplied electronics.
    Adequate provision for conversion between analog and
digital domains is therefore crucial to the utility of a
laboratory data system. Analysis of an analog/digital interface requirement will fall into roughly four categories:
    1. Signal Conditioning: The extreme variety of sensing,
measurement and control devices in use diminishes the likelihood
of direct compatibility between transducer and converter. In
practice, some forms of signal conditioning (such as amplification, filtering, ground isolation, or current to voltage to
current conversion) are frequently necessary. Because of the
application-specific nature of signal conditioning requirements,
it is difficult to provide detailed support for them in the
basic system design beyond the presence of standard regulated
supply voltages and flexible circuitry packaging provisions.
These features are present in the UNC design.
    2. Accuracy and Resolution: Absolute accuracy of conver-
sion requirements of laboratory instrumentation are usually
modest and lie in the range of one-tenth percent to one percent
of reading. Demands upon conversion resolution, however,
particularly as it relates to acceptable signal dynamic range
and conversion linearity, are often greater and may approach
one part-per-million of full scale. Advantage can be taken,
therefore, of conversion techniques such as dual-slope integration in which resolution is not inherently tied to, and limited
by, absolute accuracy.
    3. Speed: Throughput requirements of conversion hardware
may easily span six orders of magnitude with anywhere from
several data taken per minute to hundreds of thousands taken
per second.
    4. Specialized Requirements: Particular applications will
sometimes benefit from the selection of conversion techniques
with specific properties. Instrumental signals plagued by 60 Hz
related noise, for example, might be rescued by the normal-mode

rejection of a conversion technique involving integration for a
precise multiple of 1/60 second.

Analog/digital conversion is provided for in the UNC design
by two quite different modules. One, based upon the successive-
approximation technique, accepts any one of three different 12
bit ADC circuits, each available from a number of vendors. By
proper choice of ADC, maximum sample throughputs of 47,000 to
330,000 12 bit conversions per second (under DMA) are available.
Sixteen channels may be controlled. The second module is based
upon dual-slope integration and has a maximum sample rate of 20
per second. Resolution, however, is 20 parts per million of
full scale and excellent rejection of 60 Hz related noise
results from a quartz-derived integration interval of 16.6665
milliseconds. Three independently scaled channels are supported.

Digital/analog conversion in the UNC system is performed by
one or more DAC modules, each supporting up to four independent
12 bit DAC circuits. Output ranges are strappable for +5 to ±
10 volts with DMA update times of three (3) microseconds and
settling time of three (3) microseconds.

B. Time Resolution:

Reference of experimental events to time in an accurate,
precise, and jitter-free manner is a critical responsibility of
the instrumentation interface. In moderate performance
applications, it is sufficient to provide a CPU accessible
clock register or source of periodic interrupts. In the UNC
design, this role is served by a 20.000 Hz free-running eight
bit counter incorporated on the integrating ADC module capable
of interrupting the processor each 50 milliseconds.

More demaning applications cannot tolerate delays introduced by software execution, however, and can only be served by
hardware which bypasses the CPU. In the UNC design, a general
purpose I/O programmer module has been devised which manages
complex acquisition and control sequences under DMA. Time
delays are resolved to 500 nanoseconds.

C. Miscellaneous One-of-a Kind Functions:

Just as analog signal processing requirements will often
fall outside the provisions of even a broad assortment of
standard modules, so must other forms of unusual interface
requirements be accommodated. Examples within our experience
include AC power control, non-standard digital output from
instrumentation, and many others.

II. Instrumentation Controller/Local Data Processor

Because of the I/O intensive nature of laboratory data
systems, it's logical to adopt a design for the local processor
which offers maximum capability for controller modules with
minimum controller complexity. Support for efficient, high
performance instrument I/O may take the form of an I/O bus with
these properties:
   A. Simple bus structure
   B. Easily used, short response DMA
   C. Simple Interrupt Structure
   D.' Easily Decoded Device Addressing

In the UNC implementation, a 44 conductor bus was adopted
which possesses:
   A. Flexible unified bus for both I/O and memory.
   B. Simple interrupt organization.
   C. Transparent DMA at 667 K Bytes/second. Address and
       control logic on CPU and shared among peripherals.
   D. Decode of any eight-bit device address with single SSI
       device.

III. Human Operator/Processor Interface

Effective communication between the laboratory data system
and its operator depends upon efficiency of the data system in
three functional roles.

   A. Operational Dialog: Adequate interaction between
human and machine in the set up, initiation, and monitoring of
the experimental process depends upon the generation and display
by the data system of operator prompts and instrument status.
The form of operation commands accepted by the system should be
concise, free form, forgiving, and intuitively meaningful. In
the UNC design, much effort has been expended in the implementation of a library of operator dialog utilities. These utilize
the flexible, high text capacity TV display built into the UNC
hardware design and ease the programming of convenient, interactive packages.

   B. Real Time Display of Acquired Data: No form of operator feedback matches the value and assurance provided by the
immediate display of data as it is acquired. The UNC design
makes intensive use of moderate resolution (256x256) graphical
CRT display of data during or immediately after an experiment.

   C. Off-line Data Reduction: The prompt reduction and
report of experimental results with graphical capability is an
important part of the operator/processor interface. System
design should allow for operator interaction in data manipulation, along with easy programming of sophisticated data reduction algorithms.

The UNC design provides moderate resolution softcopy
graphics and high resolution annotated hardcopy graphics utilizing standard laboratory analog XY recorders. Implementation
of data reduction algorithms may currently be done either in
BASIC or in assembly language supported by a large library of
floating point math and display formatting subroutines.

IV. Mass Storage Medium/Local Processor
    Most, if not all, data system applications require the
availability of reliable, removable medium, mass storage for
both data and programs. Performance requirements dictate media
capacities which may range up to 10^6 bytes. Transfer rates
should be high compared to average data production rates of
instrumentation so that system availability is not significantly
reduced by long periods of data recording.
    An important benefit of standardized removable media mass
storage is the resulting possibility of data analysis and report
generation at times and on data systems other than those at
which experiments are performed. By this obvious strategem,
one can avoid the loss of use of expensive instrumentation
merely because the data systems upon which they depend are tied
up.
    In the UNC implementation, three mass media, DC300A tape
cartridge, phillips cassette and high density floppy disks are
currently available. All offer named file access and data
rate-medium.capacities of: Audio cassette, 200 bytes/sec -
540K bytes (one side of C90 cassette); DC300A, 6000 bytes/sec-
2.8 M bytes; Floppy Disk, 52 Kbyte/sec-600 K byte. In addition
to utility as data storage and transfer media, each device can
serve as source for automatic initial program loading (bootstrap).

V. Remote Computational Resource/Local Processor
    A frequent component of an integrated automated laboratory
is a communication network permitting transfer of data from
instrument-captive computers to a central data management and
computation resource. A variety of communication techniques
have been implemented using the UNC design. Most commonly used,
however, is straightforward asynchronous ASCII.

VI. Application Programmer/Local Processor
    An important characteristic of the application environment
of the general purpose laboratory data system is its fluid
nature. Rare is the application whose requirements are wholly

static. As a result, a typical data system must also serve as
a software development station and thus provide interface with
the application programmer.
     System properties needed to effectively implement this
software development interface are a strong function of the
language type adopted for the development effort. Generally,
development system requirements are minimized by the use of
interpretive languages such as BASIC, are enlarged if efficient
assembly language work must be supported, and are made greatest
by the choice of compiler languages like FORTRAN. The availability of a variety of mass devices, large random-access
memories, and versatile CRT display in the UNC system, together
with a large library of software development utilities, do much
to support the activities of the applications programmer.

Summary
     The foregoing text has attempted to outline the task of
the designer of general purpose data systems from the vantage
given by a form of input/output analysis, together with the
concrete example of one system design derived from such an
analysis. In the belief that the validity of the proposed
design procedure may be demonstrated by the breadth of applications served by the resulting product, a partial list of
instrumentation systems currently in use and served by UNC
microcomputers is exhibited below.
     Auger/ESCA/UPS surface analysis
     Electrochemical Analysis
     Electrophoresis Gel Scanning
     Enzyme Kinetics
     Fourier Transform NMR
     Gas Chromatography (with autosampler control)
     High Performance Liquid Chromatography (with autosampler
        and gradient control)
     Thin Layer Chromatography

References

General Purpose Microcomputers in Laboratory Automation. W.
Stephen Woodward and Charles N. Reilley. Pure & Applied
Chemistry, Vol. 50, pp. 785-799. Pergamon Press Ltd. 1978.

Use of the UNC Microcomputer in an Industrial Analytical
Environment: HPLC Applications. J. P. Koontz and S. A. Benezra.
Sept. 1979 ACS Meeting, Washington, D.C.

Circuit Simulation and 3-Dimensional Plotting in BASIC with a
Graphics Oriented Microcomputer. T. H. Ridgway, J. P. Anderson,
G. Raymond Miller, and M. C. Hill. Ibid.

Architecture and Implementation of a Simple
Multi-microprocessor

1. Introduction

   Chemists of all persuasion make extensive use of computers in
research and education. The chemists with perhaps the largest
demand for computational power are the theoretical chemists --
quantum chemists, statistical mechanicians, etc. Computational
chemistry, defined here as the a priori computational solution
to chemical problems, is now a semi-mature field, using many
well-defined methods and having, certainly, a well-defined
demand for computational resources. These demands are not being
effectively met. A few computational chemists have been fortu-
nate in obtaining large amounts of inexpensive or "free" computer
time, but many chemists, forced out of academic computer centers
by either high costs or a multitude of time-sharing users, have
been unable to apply their computational expertise in an effective fashion or contribute to the solution of significant chemical problems. There has been two recent responses to this solution. The first is the formation of the National Resource for
Computation in Chemistry (NRCC) which, as part of its initially
designated function, is supplying modest amounts of computer time
to chemists throughout the country. The second response involves
the acquisition by chemists and chemistry departments of super-
minicomputers such as Digital Equipment Corporation's VAX-11/780.
This paper, in effect,suggests a third parallel response -- that
computational chemists acquire the expertise in hardware, software, and computer architecture to design and build their own computers. This would allow them to perform their numerical experiments (calculations) at the leading edge of technology, rather

than at the whims of computer manufacturers, computer center
directors, etc.
    Computational chemistry is too established and important a
field to leave its only significant piece of apparatus to dis-
interested computer scientists. Chemists in other fields have
commonly built very sophisticated instruments -- molecular beam
machines, electron scattering spectrometers, ion cyclotron resonance spectrometers, etc. Physicists have built multi-million
(billion?) dollar accelerators. Most of today's analytical
chemists are accomplished in digital design. If computational
chemists are to take full advantage of the current revolution in
microelectonics, which offers great prospects for cost-effective high performance computation, it will be necessary to become
involved in the design and building of highly parallel architectures specific to specific chemical applications. It is not
likely that highly parallel machines will enjoy the same general
purpose nature that is characteristic of present serial machines.
Nor is it likely that there will be a sudden renewed interest
in numerical computation among computer scientists.
    Because of finite gate delays and other inherent limitations
in the speed of serial computer, high performance machines of
the future will involve a large degree of parallelism. One of
the most cost-effective ways of obtaining high performance
is likely to be a multiprocessor [1] incorporating hundreds or
possibly even thousands of microprocessors, each executing in
parallel a portion of a single program. The cost-effectiveness
of mass-produced microprocessors is the driving force behind
experimental one-of-a-kind multiprocessors such as Cm* (fifty
LSI-11 microcomputers in parallel) [2]. Many current microprocessors have rather limited and awkward instruction sets
and are totally deficient in their floating point operations.
Semiconductor technology is advancing rapidly, however, and
these deficiencies will not remain. The 16 bit NS16000 microprocessor, which will probably be announced by National Semiconductor Corporation in early 1980, may be capable of executing
a 64 bit floating point multiply in 5-10 microseconds [3]. This
speed is in the range of many present mainframe computers.
Incorporating such a microprocessor into the design of our
present system would offer the possibility of a performance
superior to that of a large mainframe (for a few specific problems) at a hardware cost under $10,000. The design of the
system described here, which employs the Zilog Z-80 microprocessor [4], is one small step in these directions.
    In this paper, we first briefly describe MICRO/1, a simple
Z-80 personal computer based on the S-100 bus. This monopro-
cessor defines the origin for our multiprocessor design. We

describe standard parallel architectures, prior to describing the specific architecture and general features of our prototype multi-processor MICRO/2. A few of the details of the actual implementation are then described. Finally, we indicate some of the software considerations enabling communication between processors, prior to presenting the conclusions of this work.

2. MICRO/1

This project began with the purchase of a personal computer (Alpha-1) produced by defunct Technical Design Labs, Inc., which subsequently became the now defunct Xitan, Inc. This is a micro-computer which uses the standard hobbyist S-100 bus [5,6]. It was built from kit form. The current configuration, which we call MICRO/1, consists of a Z-80 processor, 32K bytes of random access memory (RAM), two serial ports (RS-232) which communicate with a modified surplus teletype console device and, occasionally, a borrowed CRT terminal, an unused parallel port, and a cassette interface for loading and storing programs and data. The cost of the system, without peripherals, is in the vicinity of $1000. The cassette interface will shortly be replaced by a floppy disk system.

The software includes a 2K byte monitor held in read only memory (ROM), a text editor, macro assembler, BASIC interpreter, and application programs. Since the Z-80 is only an 8 bit processor without integer multiplication or division or floating point operations, and since BASIC interprets rather than compiles code, application programs obviously run rather slowly. Nevertheless, it is possible to perform quantum chemical calculations of educational or illustrative nature. For example, The floating spherical Gaussian orbital (FSGO) method of Frost [7] has been implemented. The program occupies about 5K bytes plus array storage (dependent on the size of molecule treated). Calculations (without optimization of the position or exponent of the orbitals) on the 10-electron molecules CH_4, NH_3, H_2O, and FH require 7 minutes, while calculations on the 18-electron molecule C_2H_6 require 59 minutes. Huckel or semi-empirical calculations (CNDO, etc.) would run effectively on a microcomputer. Ab initio self-consistent-field (SCF calculations that use only Gaussian s orbitals would be fairly easy to implement and, with the proper expertise and perhaps slightly more memory, STO-3G [8] calculations could be illustrated on a simple microcomputer such as MICRO/1.

The present configuration of MICRO/1, while simple and slow, contains the basic ingredients of much faster systems that will become available in the near future. To obtain very high performance and retain the full benefit of the low cost of microprocessors, it will be necessary to have an architecture which incorporates a large number of microprocessors working in concert on the solution of a single task. The best way of interconnecting processors and solving the associated communication problems is very much a topic of current research. The basic configuration of MICRO/1 has a number of empty slots on the S-100 bus and provides a starting point for exploring simple interconnecting (switching) structures.

3. Parallel Architectures

The modularity and low cost of new semiconductor products allow a high degree of freedom to designers of parallel systems. There may be as many ways of incorporating parallelism in an architecture as there are individuals interested in the problem. The usual way of describing parallelism [9] is in terms of the instruction stream or the data stream. The usual serial computer has a single control unit issuing a single instruction stream to a single processing unit which operates on a single data stream. It is thus a single instruction stream, single data stream (SISD) machine. A single instruction stream, multiple data stream (SIMD) machine, such as ILLIAC IV [10], has a single control unit but multiple processing units executing the same instruction with different data. This architecture allows the simultaneous broadcasting of an instruction to all processing units. The most general architecture, in this classification scheme, is that of the multiple instruction stream, multiple data stream (MIMD) machine. With this architecture, one has multiple processors (the combination of a control unit and processing unit). If, at one extreme, the processors are loosely coupled and comminicate by low-speed serial lines, the MIMD architecture is termed a network. If, at the other extreme, the processors are tightly coupled and communicate via a high-speed common global memory, the MIMD architecture is termed a multiprocessor.
The switching structure connecting processors and memory in a multiprocessor may take a number of forms. The simplest is the single bus structure in which all processors and memories attach to, and communicate via, the same bus. If the communication rate is high, performance will be severly degraded by saturation of the bus bandwidth. The crossbar switch of C.mmp [11] and the hierarchical switching structure of Cm* [2] allow much higher communication rates. Communication is the principal determinant of multiprocessor performance.

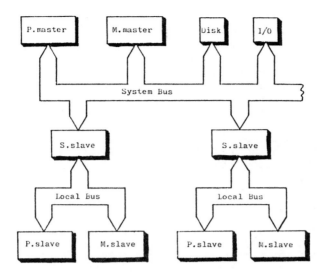

Figure 1.   General Architecture of MICRO/2.

4. MICRO/2: General Features

The architecture of MICRO/2 is shown in Figure 1 using the PMS (P=processor, M=memory, S=switch) notation of Bell and Newell [12]. The proposed architecture is not specific to the MICRO/2 implementation and we will first discuss its general characteristics and then indicate a few of the details of our specific MICRO/2 implementation in the next section.

Unlike nor democratic multiprocessors, the architecture of Figure 1 has fixed master-slave hierarchy. The master is essentially identical to any stand-alone single-bus computer. However, any number of slave modules can be attached to the system bus in the same way that peripherals are attached. Each slave module consists of a slave processor and its associated slave memory, connected by a local bus. In addition, each slave has a switch which can either completely disconnect a slave from the system bus, isolating it for local computation, or (under conditions which will shortly be described) the switch can connect the local bus to the system bus for master-slave communication. The abstract multiprocessor allows any processor in the system to access any memory module in the system. This architecture does not allow such general communication. A slave processor can only access its own local memory. Normal operation is such that all switches are "open", disconnecting each slave from the system bus, and all slaves, as well as the master, execute their own programs in complete isolation. Communication, when required, comes about as follows: Each slave switch has a register which is in the address space of the master. By writing to these memory locations, the master has complete control over any slave. The master, by writing specific bits into a switch register, may stop, continue, or restart the execution of any slave's program. In addition, the master, with the assistance of the switch, may perform direct memory access (DMA) operations. That is, it may suspend a slave processor and cause the slave to disconnect itself from the local bus, placing its address lines, data lines, etc. in the high impedance state. The local bus will then be connected to the system bus and the master can read or write to a slave memory just as it reads or writes to its own master memory (the addresses of a slave memory are disjoint from the addresses of the master memory). In this way, the master can read or write data, programs, messages, etc. to a slaves memory and completely control a slave's operation.

As the above discussion shows, the master has complete control over a slave's operation. For a completely flexible system, however, there must be some way for a slave to signal the master that it has completed a calculation, requires more data, has encountered an error, etc. This is accomplished by an interrupt mechanism. The switch register is also in the address space of a slave. By setting a specific bit of the switch register, a slave can cause an interrupt line to the master to become active. This interrupt line is a "wired-or" line, i.e. it becomes active if any of the slaves require attention. When the master is ready to service slaves, it enables interrupts, waits for a slave to interrupt, and then polls the slaves (reads the switch registers in a round-robin fashion) to determine which slaves need attention. The nature of the attention request by a slave can be determined by a message left by the slave in its local memory. In the same way, the master may leave messages for the slaves. The protocol for these message transactions is determined by system software or the specific algorithm being solved, independent of the hardware architecture.

A slightly more detailed diagram of a slave is shown in Figure 2. As indicated in the figure, registers in the switch are accessible from either the local bus (slave) or the system bus (master). The individual bits of these registers activate

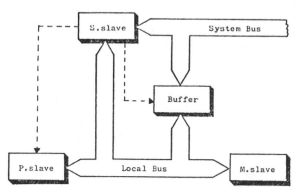

Figure 2.   Block diagram of a slave.

control lines (dotted lines in the Figure), to suspend, continue,
or restart the slave processor, or to enable the tri-state buffer.
The buffer is normally disabled (in the high impedance state),
isolating the local bus from the system bus. Under control of
the switch, however, the buffer can be enabled (subsequent to
suspending the slave processor and ensuring that the slave processor's lines are in the high impedance state) to connect the
system bus with the local bus, for DMA access of the slave memory by the master. The buffer also translates system bus signals
to the possibly slightly different signals of the local bus.

   The proposed architecture has one other extremely important
feature -- a broadcast mode. Each switch is constructed so that
it answers not only to its own specific address (on a read or
write request from the master) but also to a broadcast address
that is identical for all slave switches. In addition, each
slave memory occupies the identical position in the address
space of the master. The master can therefore broadcast instructions, data, programs, etc. to all slaves simultaneously. There
must, of course, be a hardware or software facility disallowing
simultaneous reading of each slave memory as this would lead to
garbled information.

   The architecture that we have described is not suitable for
parallel algorithms that require a very high rate of communication between arbitrary processors. Nevertheless, it could be an
extremely effective architecture for certain problems. It is a
very simple architecture. The complexity of the switch is such
that it could exist on a single chip, very unlike the complex

crossbar switch or Kmap switch of the large multiprocessors
C.mmp [11] and Cm* [2]. The design is essentially one in which
a master passes out processes (subdivisions of a single task) to
slave processors and collects the results. The broadcast mode
allows rapid transfer of identical routines and/or data to all
slaves. Perhaps the simplest example of the use of such an architecture would be the classical trajectory calculations of chemical kinetics. The slaves would calculate independent trajec-
tories and the master would simply collect the results for statistical averaging and the generation of a reaction rate constant. It appears to the authors that the architecture would
also be very effective in the molecular orbital calculations of
quantum chemistry and the molecular dynamics simulations of
statistical mechanics.

5. MICRO/2: Implementation

In the context of this article it is not possible to completely specify MICRO/2, our implementation of the above architecture. However, sufficient details will be given so that the
flavor of the design can be appreciated. MICRO/1 is an S-100
computer with a number of empty bus slots. A complete slave
module (switch, buffer, processor, and 16K bytes of memory will
fit on a single wire-wrapped S-100 PC card and can just be inserted into an existing card cage. Converting MICRO/1 to a multiprocessor thus consists only of adding one or more slave cards
to S-100 bus slots. The cost of a slave module is in the vicinity of $300.
    The master memory occupies the addresses 0000_{16} to $BFFF_{16}$
(48K bytes, although only 32K bytes are presently being used).
Each 16K byte slave memory occupies the adjacent addresses
$C000_{16}$ to $FFFF_{16}$. The master thus addresses either its own
memory or a slave memory depending on whether the address it
sends out occurs in the lower 3/4 or upper 1/4 of its total
address space. To access a slave memory, that slave's switch
must first have been requested to suspend the slave processor
and enable the buffer connecting the local bus to the system
bus. The only alteration that has needed to be made to our
master (MICRO/1) involves its monitor, which resides in ROM at
$F000_{16}$ to $F777_{16}$. A relatively simple fix disables this ROM
when the master accesses a slave memory. If the monitor had
resided in the lower 3/4 of the address space, no hardware
alteration, whatever, would be required of MICRO/1 in order for it
to serve as a master for MICRO/2.

The switch registers have an address in the I/O space of the Z-80 rather than in the memory space. Of the 256 I/O addresses, the first 128 (00_{16} to $7F_{16}$) are allocated to I/O devices, the next 64 (80_{16} to BF_{16}) are allocated to slave switch registers, and any I/O address in the range $C0_{16}$ to FF_{16} specifies a broadcast to all switch registers (there is no reason to com-pletely decode the broadcast address). Although these addresses allow up to 64 slaves, the power supply, cabinet, etc. would have to be redesigned to accomodate this many.

A block diagram of a slave, showing the address, data, and control lines, is given in Figure 3. The S-100 signals SINP, SOUT, and the address and data lines are for reading and writing the switch register by the master. The $\overline{PRESET}$ line, leading to the $\overline{RESET}$ line of the Z-80 processor, forces a reset of the system on power up and on pushing the reset button. The $\overline{PINT}$ line is the interrupt line leading to the master; it becomes active when a slave requests attention. The S-100 clock signal ϕ_1 is identical to the Z-80 clock pulse ϕ. The S-100 write signal $\overline{PWR}$ is translated without change to the Z-80 signal $\overline{WR}$. The S-100 signals SINP, SOUT, and PSYNC are combined to form the Z-80 signal $\overline{MREQ}$. Our design uses dynamic memory, although this is not required or necessarily even desired, and special circuitry associated with the memory allows refresh ($\overline{RFSH}$) to come from either the slave or, during DMA, the master.

Since a slave has no peripheral units, we have implemented the signal which requests attention of the master as the I/O request line ($\overline{IORQ}$). Whenever a slave executes an input or output instruction, this line becomes active, setting an atten-tion bit in the switch register and causing the $\overline{PINT}$ line to become and remain active until the bit is reset by the master. Using interrupt mode 1, a Z-80 master processor will vector to the fixed location 0038_{16} when interrupted by a slave. DMA references occur by the master setting a DMA bit, which results in the bus request line ($\overline{BUSRQ}$) to the slave processor becoming active. After completing its current instruction, the Z-80 slave sets its bus lines in the high impedance state, discon-necting itself from the local bus, and activates the bus acknow-ledge line ($\overline{BUSAK}$). This eventually results in the DMA line becoming active, enabline the tri-state buffer. Requesting a suspended processor to continue simply amounts to clearing the DMA bit. The switch contains 3 bits which may be set or cleared by the master -- the attention bit, the DMA bit, and a start bit. Only the attention bit can be set by the slave. The start bit causes a restart of the slave processor.

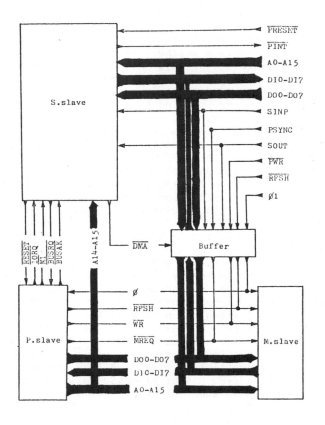

Figure 3.   Address, data, and control lines of a MICRO/2 slave.

On restart, a Z-80 processor always begins executing instructions at address 0000_{16}. To force the slave to begin at $C000_{16}$,
special circuitry in the switch clamps the data lines low,
causing the processor to execute NOPS (null instructions) until
address $C000_{16}$ is reached. Address lines A_{14}-A_{15} and the
instruction fetch signal ($\overline{M1}$), leading from the slave processor
to the switch, are for this purpose.
    As an illustration of part of the circuitry in the switch,
Figure 4 shows the explicit details of the attention circuitry.
The attention bit is a single D-type flip-flop. The bit is
set by the $\overline{IORQ}$ line from a slave and cleared, on initialization, by the $\overline{PRESET}$ line. The complemented output of the
flip-flop pulls the interrupt line down when the flip-flop
is in the set state. The diode isolates the switch from
others which may be pulling the interrupt line low. The
address decoding and SINP and SOUT lines cause a read ($\overline{RD}$) or
a write (WR) line to become active when the master
requests an input or output to this switch.

6. Software Considerations

    The simple architecture that we have described is quite
flexible. The ways in which the hardware can be used depend
very much on the software that is implemented. As an example
of such software, we illustrate here a simple slave protocol
for requesting service from the master. Analogous (but more
complex) protocols exist for the operations of the master.
The Z-80 code for a slave is shown in Figure 5. Upon restart,
a slave will enter at address $C000_{16}$ and immediately jump to a
user program. If at any time the slave would like attention
from the master -- to deliver results to the master, for example,
the slave calls the routine at label "attn". At this address,
prior to setting the attention bit by executing any "out"
instruction, the slave sets a flag to 1, to indicate that it
has not yet been serviced (upon completing the servicing of a
slave, the master will set this flag to 0). The slave then
continues any calculations that do not depend on the servicing.
When the slave can no longer continue without having been
serviced, it calls the routine at label "wait". If the flag has
been set to 0 by the master, then the slave knows it has been
serviced and returns to its calculations. If the flag still has
a value of 1, the slave executes a wait loop until the value
becomes 0 and servicing must have occurred.

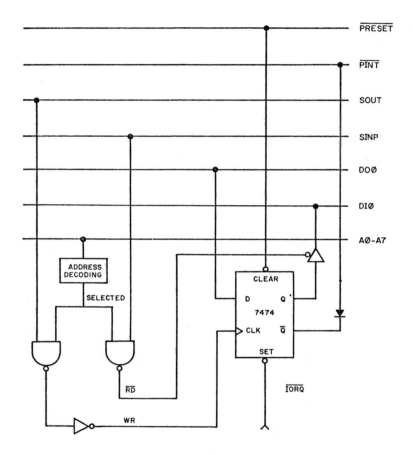

Figure 4. The attention circuitry of a MICRO/2 switch

```
CØØØ                .loc    CØØh    ; start of slave memory
            ;
CØØØ        entry:  jmp     start   ; go to user program
CØØ3        flag:   .byte   ØØh     ; Ø=continue, 1=wait
            ;
            ; Signal for attention and continue execution
            ;
CØØ4        attn:   mvi     a,Ø1h   ; set flag=1
CØØ6                sta     flag    ; for wait
CØØ9                out     ØØh     ; signal for attention
CØØB                ret             ; continue executing
            ;
            ; Wait for master if necessary
            ;
CØØC        wait:   lda     flag    ; test flag to see
CØØF                cpi     ØØh     ; if master came
CØ11                jrnz    wait    ; wait for master
CØ13                ret             ; service completed, continue
            ;
CØ14        start:                  ; user program starts here
                    .end
```

Figure 5. Communication protocol for a slave.

7. Conclusion

We have described here a simple architecture for a new multiprocessor and a specific Z-80, S-100 implementation of that architecture. The general architecture shows promise of providing high performance (and extremely cost-effective performance) for particular applications. The implementation, which we term MICRO/2, is an excellent tool for experimenting with design problems involving the architecture and for exploring the applicability of the architecture to parallel algorithms for some of the common problems of computational chemistry. The final prototype version, which will employ two slave modules, is not yet completely assembled, but a breadboard version of the switch behaves as expected. Programming and experimentation with parallel algorithms will begin in the near future.

The Z-80, S-100 version of MICRO/2 is limited in its performance mainly as a result of the inability of the Z-80 to execute floating point operations and integer multiplication and division. These deficiencies in current microprocessors will rapidly disappear. Once sufficient experience is gained with a working version of the current prototype design, efforts will begin on the design of a similar system around one of the new 16 bit microprocessors and an associated floating point chip. The definition of the S-100 bus has recently been extended [13] to include 16 bits of data and 24 bits of address. Even within the simple S-100 framework, a "personal" multiprocessor of a few years hence will be capable of rapid execution of scientific "number-crunching" problems.

References

[1]   see N. S. Ostlund, Intern. J. Quantum Chem. S13, 15 (1979) for a discussion of the use of multiprocessors for chemical calculations.

[2]   S. H. Fuller, J. K. Ousterhout, L. Raskin, P. I. Rubinfeld, P. J. Sindhu, and R. J. Swan, Proc. IEEE, 66, 216 (1978).

[3]   personal communication from National Semiconductor Corporation.

[4]   Zilog, Inc. 10460 Bubb Road, Cupertino, California 95014.

[5]   W. M. Goble, Interface Age, 2, 66 (1977), June.

[6]   K. A. Elmquist, Interface Age, 3, 122 (1978), August.

[7]   A. A. Frost, B. H. Prentice, and R. A. Rouse, J. Amer. Chem. Soc., 89, 3064 (1967).

[8]   W. J. Hehre, R. F. Stewart, and J. A. Pople, J. Chem. Phys., 51, 2657 (1969).

[9]   M. Flynn, IEEE Trans. Computers, C-21, 948 (1972).

[10]   G. H. Barnes, R. M. Brown, M. Kato, D. J. Kuck, D. L. Slotnick, and R. A. Stokes, IEEE Trans. Computers, C-17, 746 (1968).

[11]   W. A. Wulf and C. G. Bell, AFIPS Conf. Proc., 41, 765 (1972).

[12]   C. G. Bell and A. Newell, "Computer Structures: Reading and Examples", McGraw Hill, New York, 1971.

[13]   K. A. Elmquist, H. Fullmer, D. B. Gustavson, and G. Morrow, Computer, 12, 28 (1979).

INDEX

Accuracy and reso-
lution, 234
Adaption circuits, 61
Agriculture products,
92
Altair 880, 13
Altair 8800b, 60
Analogrates of 100
MHz, 13
Analog signals, 61
Analog to digital
convertor A/D, 49
APL, 224
Apple II, 189, 199
Assembler, 170
Autoinjector, 3, 4
Automated analysis,
10
Autoranging, 29

Autosampler, micro-
processor-based, 1
Autoscaling, 2
Averaging, moving
window, 55

Base line, 1
automatic, 18
Base line analysis,
24
BASIC, 3
DISK, 49
disk operating
system (BDOS),
213
5K Processor
Technology, 16
I/O System (BIOS),
213

Program commons, 143
Program translators,
   213
Program transport-
   ability, 210
Prompt, 140
Protocol, 248
Proton NMR, 180
Psuedo code (p-code),
   217
Pulse radiolysis, 12

Quadrature, numerical
   integration, 156
Quality assurance, 9
Quantum chemical
   calculations, 243

Radio Shack TRS-80,
   189
Random access, 210
Raw data storage and
   retrieval, 2
Real time clock, 119
Relay, optically
   isolated, 39
Reliability, 106
Repetitive scanning,
   110
Reproducibility,
   78
Retention, relative,
   79
Retention time, one
   peak, 79
   two peak, 79
Router index, 181
RS-232-C bit serial
   interface, 28
Run editors, 213
Runge-Kutta procedure,
   160

Sampling rate, 68
Satellite micro-
   computer, 86
S-100 bus, 13, 22,
   208
   IEEE standard, 209
Screen-oriented
   editor, 226
Sense line, decimal
   point, 119
Sense lines, 120
Serial data channels,
   210
Signal averager, 58
Signal averaging, 65,
   126
Signal conditioning,
   234
Simulation of experi-
   ments, populations
   or environments,
   193
Single instruction
   single data
   stream (SISD),
   242
Smoothing, 19
Soft return, 144
Soft sector recording
   format, 210
Software development,
   238
Solvent scouting,
   9
Source code, 217
Space compression, 230
Specific activity, 100
Spectrometer, IR, 73
   UV-visible, 73
Spectrophotometer,
   Cary 17N, BCD,
   105

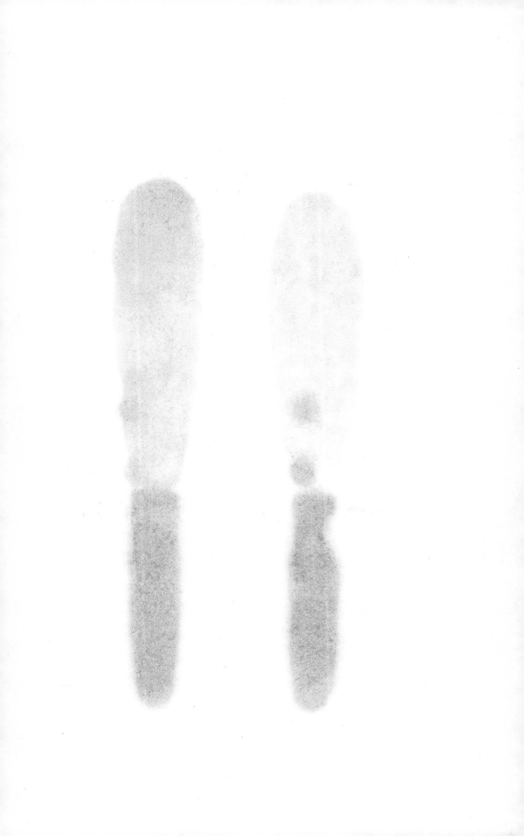

DATE DUE

DE UY '93 8/29/99 11-11			